NES Essential Academic Skills Math: 5 NES Math (003) Practice Tests with 225 Questions and Solutions

NES and National Evaluation Systems are registered trademarks of Pearson, Inc, which is not affiliated with nor endorses this publication.

NES Essential Academic Skills Math: 5 NES Math (003) Practice Tests with 225 Questions and Solutions

© COPYRIGHT 2017 Exam SAM Study Aids & Media

All rights reserved. No part of this publication may be reproduced, stored in a retrieval system, or transmitted, in any form or by any means, electronic, mechanical, photocopying, recording, or otherwise, without the prior written permission of the copyright owner.

ISBN-13: 978-1-949282-19-1
ISBN-10: 1-949282-19-8

For information on bulk discounts, please contact us at: email@examsam.com

The drawings in this publication are for illustration purposes only. They are not drawn to an exact scale.

NOTE: NES and National Evaluation Systems are registered trademarks of Pearson, Inc, which is not affiliated with nor endorses this publication

TABLE OF CONTENTS

NES Practice Math Test 1 with Study Tips

Number properties and number operations:

Integers and Signed Numbers	1
Fractions	1
Mixed Numbers	1
PEMDAS – Order of Operations	2
Proportions and Ratios	3
Properties of Exponents	3
Scientific Notation	3

Algebra and graphing:

Evaluating expressions	3
Determining possible values for variables	4
Inequalities	4
Solving for x	4
Linear equations:	
Calculating slope	5
Calculating x and y intercepts	5
Positive and negative linear relationships	6
Determining missing values in linear equations	7
Working with functions	8

Geometry and measurement:

Angle Measurement	9
Area – Circles, Rectangles, and Triangles	9
Pythagorean Theorem and Hypotenuse Length	10
Perimeter of Squares and Rectangles	10
Circumference and Diameter	11
Volume – Rectangular Solids, Cones, and Cylinders	11
Weights and Measures	12

Probability and statistics:

Mean	12
Mode	12
Median	13
Range	13
Distribution and spread	13
Probability	14
Determining missing value from sample space	14

Understanding statistical conclusions and outcomes	14
Interpreting data from pictographs, charts, and graphs	15

Problem solving, reasoning, and mathematical communication:

Numerical sequences	15
Deductive reasoning	15
Estimating	16
Reasoning	16
Mathematical communication	17
Understanding data representation	17
Working with algorithms	18
Answer Key for NES Practice Math Test 1	19
NES Practice Math Test 1 – Solutions and Explanations	20

NES Practice Math Test 2 with Study Tips

Number properties and number operations:

Fractions – advanced problems	32
Percentages and decimals – advanced problems	33
Operations on integers – advanced problems	34
Exponents – advanced problems	34

Algebra and graphing:

Algebraic expressions – advanced problems	35
Inequalities – advanced problems	35
Equivalent expressions	36
Practical problems	36
Polynomials	37
Linear equations – midpoints	37

Geometry and measurement:

Geometry – advanced problems	38
Perimeter – advanced problems	40
Volume – advanced problems	40
Pythagorean Theorem – advanced problems	40
Triangle laws	41
Hybrid shapes	42

Probability and statistics – advanced problems

Further practice with mean, median, mode, and range	43
Interpreting line graphs and bar charts	44
Distribution – advanced problems	46

Advanced problem solving, reasoning, and mathematical communication   47

Answer Key for NES Practice Math Test 2	51
NES Practice Math Test 2 – Solutions and Explanations	52

NES Practice Math Test 3:

Number properties and number operations	62
Algebra and graphing	63
Geometry and measurement	68
Probability and statistics	71
Problem solving, reasoning, and mathematical communication	74
Answer Key for NES Practice Math Test 3	78
NES Practice Math Test 3 – Solutions and Explanations	79

NES Practice Math Test 4:

Number properties and number operations	86
Algebra and graphing	87
Geometry and measurement	89
Probability and statistics	92
Problem solving, reasoning, and mathematical communication	96
Answer Key for NES Practice Math Test 4	99
NES Practice Math Test 4 – Solutions and Explanations	100

NES Practice Math Test 5:

Number properties and number operations	107
Algebra and graphing	108
Geometry and measurement	112
Probability and statistics	115
Problem solving, reasoning, and mathematical communication	118
Answer Key for NES Practice Math Test 5	120
NES Practice Math Test 5 – Solutions and Explanations	121

How to Use This Publication

As you work through this study guide, you will notice that Practice Tests 1 and 2 are in workbook format, providing study tips after each question.

The format of the first two practice tests introduces the exam concepts and helps you learn the strategies and formulas that you need to answer all of the types of questions on the actual NES Math Test.

You can refer back to the formulas and tips introduced in the first two practice tests as you work through the remaining material in the book.

You may wish to time yourself as you do the practice tests in this book, allowing yourself 60 minutes for each exam. This will help to simulate the conditions of the actual test.

The answers and solutions for the practice tests are provided at the end of each of the practice exams.

This study guide assumes knowledge of basic math skills, such as addition, subtraction, multiplication, division, percentages, and decimals.

If you have difficulties with basic math problems or if you have been out of school for a while, you may wish to review our free basic math problems before taking the practice tests in this book.

The free review problems can be found at: www.examsam.com

NES Essential Academic Skills Math Test Format

The NES Essential Academic Skills Mathematics Test contains the following types of questions:

- Number properties and number operations
- Algebra and graphing
- Geometry and measurement
- Probability and statistics
- Problem solving, reasoning, and mathematical communication

There are approximately 45 questions in total on the NES Essential Academic Skills math exam.

You will take the exam on a computer, unless you have applied for an exemption.

You will have an hour to take the NES Essential Academic Skills math test.

Number sense and number operations questions cover the following skills:

- Performing addition, subtraction, multiplication, and division
- Understanding place value
- Applying operations for integers, fractions, decimals, proportions, and percentages
- Using mathematical equivalents, exponents, and scientific notation

Algebra and graphing questions cover:

- Evaluating algebraic expressions
- Equations and inequalities with one variable
- Algebraic equivalents
- Linear equations
- Graphing

Geometry and measurement questions cover these skills:

- Identifying and converting units of measurement
- Solving problems with lines, line segments, angles
- Understanding the properties of geometric shapes
- Length and area
- Perimeter and circumference
- Pythagorean Theorem

Probability and statistics questions cover:

- Calculating probability
- Reading and interpreting graphs, charts, tables, and pictographs
- Understanding appropriate and inappropriate uses of statistical measures
- Calculating mean, mode, median, range, and spread
- Interpreting distributions
- Understanding positive and negative patterns in data

Problem solving, reasoning, and mathematical communication questions cover:

- Estimating solutions to problems
- Using instructions to solve problems
- Using inductive and deductive reasoning
- Understanding mathematical terminology

The questions on the NES Essential Academic Skills Mathematics Test will be multiple-choice, with answers choices A to D. For certain questions, reference materials or formulas may be made available on the screen.

NES Essential Academic Skills Practice Math Test 1 with Study Tips

Number properties and number operations:

1) $-(-5) + 3 = ?$
   A) −8
   B) −2
   C) 2
   D) 8

> Computations with signed numbers are frequently included on the NES Essential Academic Skills examination. Many of these types of problems will involve integers. Integers are positive and negative whole numbers. Integers cannot have decimals, nor can they be mixed numbers. In other words, they can't contain fractions. One of the most important concepts to remember when working with signed numbers is that two negative signs together make a positive number. So, when you see a number like − (−2) you have to use 2 in your calculation.

2) What is $1/3 \times 2/3$?
   A) $2/3$
   B) $2/6$
   C) $2/9$
   D) $1/3$

> You will see problems on the exam that ask you to multiply fractions. To multiply fractions, you first need to multiply the numerators from each fraction. Then multiply the denominators. The numerator is the number on the top of each fraction. The denominator is the number on the bottom of the fraction.

3) $3\frac{1}{3} - 2\frac{1}{2} = ?$
   A) $\frac{1}{3}$
   B) $\frac{9}{3}$
   C) $\frac{5}{6}$
   D) $1\frac{1}{2}$

> Mixed numbers are those that contain a whole number and a fraction. Convert the mixed numbers back to fractions first. Then find the lowest common denominator of the fractions in order to solve the problem.

4) $-6 \times 3 - 4 \div 2 = ?$
   A) -20
   B) -18
   C) -2
   D) 4

This question tests your knowledge of order of operations. The phrase "order of operations" means that you need to know which mathematical operation to do first when you are faced with longer problems. Remember the acronym PEMDAS. "PEMDAS" means that you have to do the mathematical operations in this order:
First: Do operations inside **P**arentheses
Second: Do operations with **E**xponents
Third: Perform **M**ultiplication and **D**ivision (from left to right)
Last: Do **A**ddition and **S**ubtraction (from left to right)

---

5) $\dfrac{5 \times (7-4)^2 + 3 \times 8}{5 - 6 \div (4-1)} = ?$
   A) -23
   B) 23
   C) $\dfrac{23}{1/3}$
   D) 46

This is an advanced question on order of operations. Remember PEMDAS:

Parentheses – Exponents – Multiplication & Division – Addition & Subtraction

---

6) Find the value of x that solves the following proportion: $3/6 = x/14$
   A) 3
   B) 6
   C) 7
   D) 8

A proportion is an equation with a ratio on each side. In other words, a proportion is a statement that two ratios are equal. $3/4 = 6/8$ is an example of a proportion. We will look at ratios again in the next question.

---

7) In a shipment of 100 mp3 players, 1% are faulty. What is the ratio of non-faulty mp3 players to faulty mp3 players?
   A) 1:100
   B) 100:1
   C) 99:100
   D) 99:1

Ratios take a group of people or things and divide them into two parts. For example, if your teacher tells you that each day you should spend two hours studying math for every hour that you spend studying English, you get the ratio 2:1. The number before the colon expresses one subset of the total amount of items. The number after the colon expresses a different subset of the total. In other words, when the number before the colon and the number after the colon are added together, we have the total amount of items.

8) $11^5 \times 11^3 = ?$
A) 11^8
B) 11^{15}
C) 22^8
D) 121^8

You will need to know properties of exponents for the examination. You will see questions on the exam that involve adding and subtracting exponents. When the base numbers are the same and you need to multiply the base numbers, you add the exponents. When the base numbers are the same and you need to divide, you subtract the exponents.

9)   Express 784 in scientific notation.
A) $7840 \times 1/10$
B) $784 \times 10/{-0}$
C) 78.4×10
D) 7.84×10^2

Scientific notation means that you express a number as a digit number and a multiple of ten. The digit number has a number in the ones place and the other numbers in the decimal places. The multiple of ten is expressed as ten to a certain power. For example: $756,000,000 = 7.56 \times 10^8$

Algebra and graphing:

10)   Evaluate the expression $4x^2 + 2xy - y^2$ for $x = 2$ and $y = -2$?
A) 4
B) 6
C) 8
D) 12

You may be asked to "evaluate" expressions on the exam. This means that you have to calculate the value of the expression by substituting its values. To solve these problems, put in the numbers stated for x and y and multiply. In this question, $x = 2$ and $y = -2$. Once you have done the multiplication, do the addition and subtraction.

11)   What are two possible values of x for the following equation? $x^2 + 6x + 8 = 0$
A) 1 and 2
B) 2 and 4
C) 6 and 8
D) −2 and −4

> You may see problems on the exam that give you a quadratic equation and ask you to determine possible values for the variables in the expression. First, factor the equation into this format: $(x + \_)(x + \_)$. Then substitute 0 for one of the x's and work out an answer. Then substitute zero for the other x to work out the other possible answer.

12) In the equations below, x represents the cost of one online game and y represents the cost of one movie ticket. If $x - 2 > 5$ and $y = x - 2$, then the cost of 2 discounted movie tickets is greater than which one of the following?
    A) $x - 2$
    B) $x - 5$
    C) $y + 5$
    D) 10

> Inequality problems will have a less than or greater than sign. When solving inequality problems, isolate integers before dealing with any fractions. Also remember that if you multiply an inequality by a negative number, you have to reverse the direction of the less than or greater than sign.

13) A company sells jeans and T-shirts. J represents jeans and T represents T-shirts in these equations: $2J + T = \$50$ and $J + 2T = \$40$. Sarah buys one pair of jeans and one T-shirt. How much does she pay for her entire purchase?
    A) $10
    B) $20
    C) $30
    D) $40

> Several questions on the NES Essential Academic Skills Test will ask you to solve practical problems. Practical problems may involve calculating prices or discounts for items in a store. Other common practical problems involve calculations with exam scores or other data for a class of students.

14) Solve for x: $3x - 2(x + 5) = -8$
    A) 1
    B) 2
    C) 3
    D) 5

> You will see problems involving solving equations for an unknown variable on the exam. To solve the problem, perform the multiplication on the items in parentheses first. Then eliminate the integers by isolating them to one side of the equation. Finally, perform any other operations to solve for x.

15) Marta runs up and down a hill near her house. The measurements of the hill can be placed on a two dimensional linear graph on which $x = 5$ and $y = 165$. If the line crosses the y axis at 15, what is the slope of this hill?
    A) 10
    B) 20
    C) 30
    D) 36

Calculating slope and slope intercept are two important skills that you will need for algebra and graphing problems on the exam. To put it in simple language, slope is the measurement of how steep a straight line on a graph is. Slope will be negative when the line slants upwards to the left. On the other hand, slope will be positive when the line slants upwards to the right. The two points are represented by the coordinates (x_1, y_1) and (x_2, y_2).

Slope is represented by variable m. We can calculate slope by using the slope formula. The slope formula is as follows:

$$m = \frac{y_2 - y_1}{x_2 - x_1}$$

You will sometimes be given a set of points, and then told where the line crosses the y axis. In that case, you may also need what is known as the slope-intercept formula. In the slope-intercept formula, m is the slope, b is the y intercept (the point at which the line crosses the y axis), and x and y are points on the graph. Here is the slope-intercept formula:

$$y = mx + b$$

16) Find the x and y intercepts of the following equation: $x^2 + 4y^2 = 64$
   A) (8, 0) and (0, 4)
   B) (0, 8) and (4, 0)
   C) (4, 0) and (0, 8)
   D) (0, 4) and (8, 0)

You may also be asked to calculate x and y intercepts in algebra and graphing problems. The x intercept is the point at which a line crosses the x axis of a graph. In order for the line to cross the x axis, y must be equal to zero at that particular point of the graph. On the other hand, the y intercept is the point at which the line crosses the y axis. So, in order for the line to cross the y axis, x must be equal to zero at that particular point of the graph. For questions about x and y intercepts, substitute 0 for y in the equation provided. Then substitute 0 for x to solve the problem.

17) A mother has noticed that the more sugar her child eats, the more her child sleeps at night. Which of the following graphs best illustrates the relationship between the amount of sugar the child consumes and the child's amount of sleep?

A)
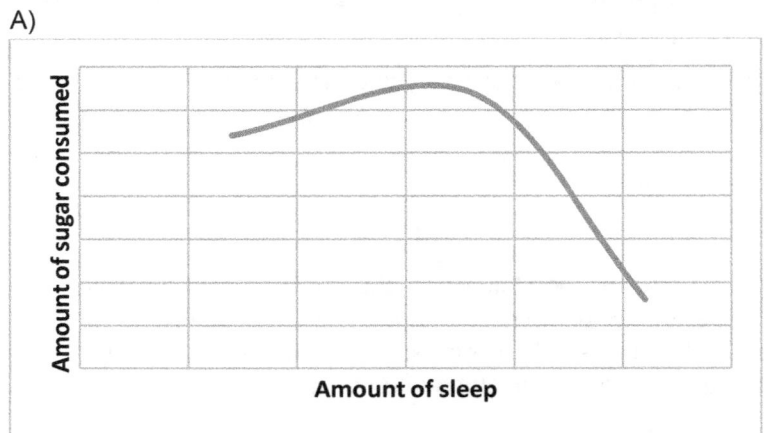

B)
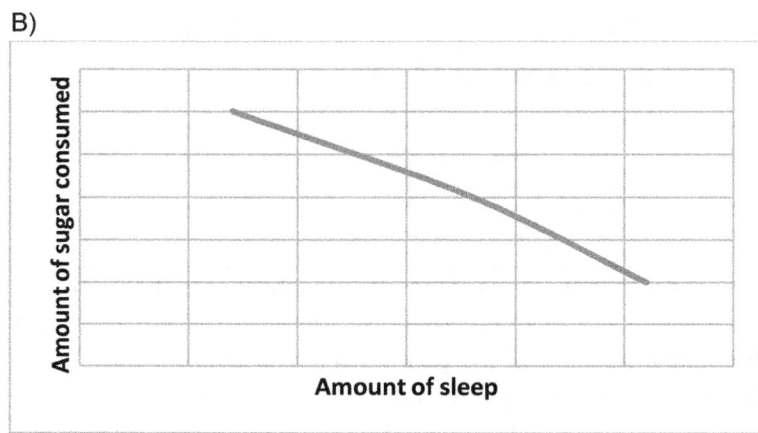

C)
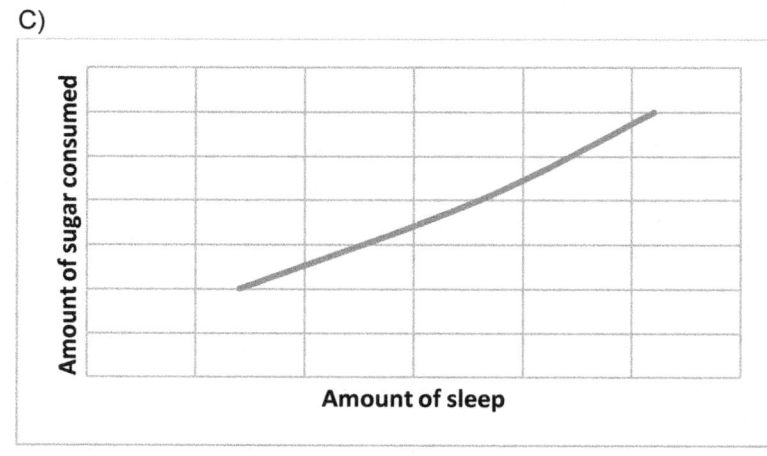

D)

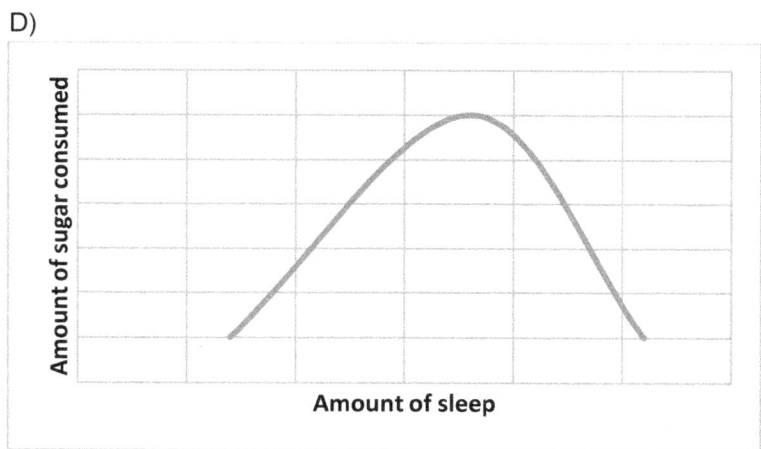

Your exam will have problems like this one that show line graphs of linear equations. Be sure that you know the difference between positive linear relationships and negative linear relationships for the exam. In a positive linear relationship, an increase in one variable causes an increase in the other variable, meaning that the line will point upwards from left to right.

In a negative linear relationship, an increase in one variable causes a decrease in the other variable, meaning that the line will point downwards from left to right.

18) The graph of a linear equation is shown below. Which one of the tables of values best represents the points on the graph?

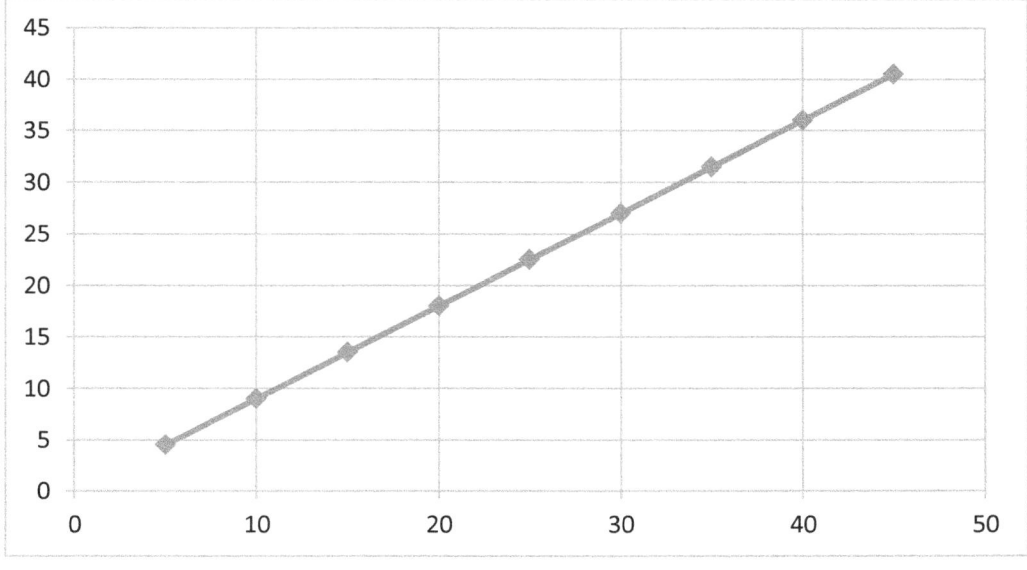

A)

x	y
5	5
10	10
15	15
20	20

B)

x	y
5	4
10	8
15	12
20	16

C)

x	y
5	4.5
10	9
15	13.5
20	18

D)

x	y
5	9
10	13
15	15
20	20

This is an example of an exam question on graphing that involves functions. A function expresses the mathematical relationship between x and y. Functions can be expressed by using the notation: $f(x)$. To solve problems like this one, try to determine what recurring mathematical operation on x will yield a result of y. In this question, we have the function: $f(x) = x \times 0.9$. For instance, $f(x) = 5 \times 0.9 = 4.5$.

Geometry and measurement:

19) Consider the isosceles triangle in the diagram below.

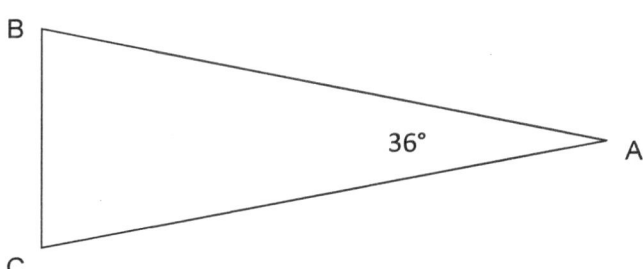

What is the measurement of ∠B?
A) 36°
B) 45°
C) 72°
D) 144°

For angle measurement questions, you need to remember these concepts: (1) The sum of all three angles in a triangle is always 180°; (2) Two sides of an isosceles triangle are equal in length, and their corresponding angles are also equal; (3) For an isosceles triangle, deduct the degrees given from 180° to find out the total degrees of the two other angles.

20) A football field is 100 yards long and 30 yards wide. What is the area of the football field in square yards?
A) 130
B) 150
C) 300
D) 3000

You will need to calculate the area of geometric shapes, such as circles, squares, triangles, and rectangles for the test. In this problem, we are working with a rectangle. Be sure that you know how to use these formulas for the exam:

Area of a circle: $\pi \times r^2$ (radius squared)

Area of a square or rectangle: length × width

Area of a triangle: (base × height) ÷ 2

21) If one side of a triangle is 5cm and the other side is 12cm, what is the measurement of the hypotenuse of the triangle?
A) $5\sqrt{12}$ cm
B) $12\sqrt{5}$ cm

C) $\sqrt{17}$ cm
D) 13 cm

The hypotenuse is the side of the triangle that is opposite the right angle. In other words, the hypotenuse is opposite the square corner of the triangle. To calculate the length of the hypotenuse in right triangles, you will need the Pythagorean Theorem. According to the theorem, the length of the hypotenuse (represented by side C) is equal to the square root of the sum of the squares of the other two sides of the triangle (represented by A and B). For any right triangle with sides A, B, and C, you need to remember this formula:

$$\text{hypotenuse length C} = \sqrt{A^2 + B^2}$$

22) In the figure below, XY is 4 inches long and XZ is 5 inches long. What is the area of triangle XYZ?

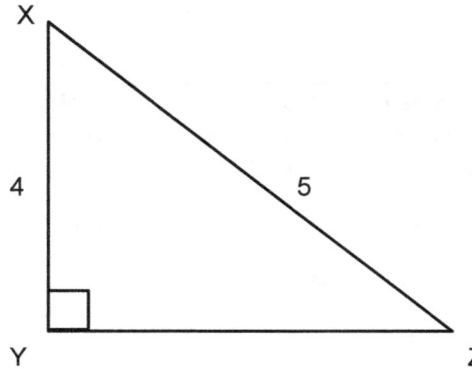

A) 3
B) 5
C) 6
D) 10

Use the Pythagorean Theorem to find the length of YZ.

Then use the formula to calculate the area of a triangle:

triangle area = (base × height) ÷ 2

23) What is the perimeter of a rectangle that has a length of 5 and a width of 3?
A) 15
B) 16
C) 18
D) 40

> The perimeter is the measurement along the outer side of a square, rectangle, or hybrid shape. In order to calculate the perimeter of squares and rectangles, you need to use the perimeter formula:
>
> $$\text{Perimeter} = (\text{length} \times 2) + (\text{width} \times 2)$$

24) If a circle has a diameter of 12, what is the circumference of the circle?
   A) 6π
   B) 12π
   C) 24π
   D) 36π

> The circumference is the measurement around the outside of a circle. You can think of circumference like perimeter, except circumference is used in calculations for circular objects, rather than for shapes like squares or rectangles. The formula for the circumference of a circle is:
>
> $$\text{Circumference} = \pi \times \text{diameter}$$

25) A box is manufactured to contain either laptop computers or tablets. When the computer systems are removed from the box, it is reused to hold other items. If the length of the box is 20cm, the width is 15cm, and the height is 25cm, what is the volume of the box?
   A) 150
   B) 300
   C) 750
   D) 7500

> The test will have questions that ask you to calculate the volume of certain geometric shapes. For the test, you may need to calculate the volume of a cylinder, cone, or rectangular solid.
>
> Rectangular solid volume: base × width × height
>
> Cone volume: $(\pi \times \text{radius}^2 \times \text{height}) \div 3$
>
> Cylinder volume: $\pi \times \text{radius}^2 \times \text{height}$

26) Which of the following is the most appropriate unit of measurement for the weight of a car?
   A) horsepower
   B) gallons
   C) pounds per square inch
   D) tons

> This is a measurement question on understanding how to measure solid and fluid weights. Be sure that you know how to use US weights and measures for the exam.

27) Find the equivalent of 6 yards in inches.
   A) 18 inches
   B) 72 inches
   C) 216 inches
   D) 1,296 inches

For questions involving conversions of measurements, take note of the unit of measurement in the calculation. For this problem, remember that 1 yard = 36 inches.

Probability and statistics:

28) The ages of 5 siblings are: 2, 5, 7, 12, and x. If the average age of the 5 siblings is 8 years old, what is the age (x) of the 5th sibling?
   A) 8
   B) 10
   C) 12
   D) 14

This is a problem on determining the value that is missing from the calculation of an average of a set of values. Remember that the mean is the same thing as the arithmetic average. In order to calculate the mean, you simply add up the values of all of the items in the set, and then divide by the number of items in the set. To solve problems like this one, set up an equation to calculate the mean, using x for the unknown value.

29) Members of a weight loss group report their individual weight loss to the group leader every week. During the week, the following amounts in pounds were reported: 1, 1, 3, 2, 4, 3, 1, 2, and 1. What is the mode of the weight loss for the group?
   A) 1 pound
   B) 2 pounds
   C) 3 pounds
   D) 4 pounds

This is a question on mode. Mode is the value that occurs most frequently in a data set. For example, if 10 students scored 85 on a test, 6 students scored 90, and 4 students scored 80, the mode score is 85.

30) Mark's record of times for the 400 meter freestyle at swim meets this season is:

   8.19, 7.59, 8.25, 7.35, 9.10

   What is the median of his times?
   A) 7.59
   B) 8.19
   C) 8.25
   D) 8.096

This question is asking you to find the median of a set of numbers. The median is the number that is in the middle of the set when the numbers are in ascending order.

31) A student receives the following scores on her assignments during the term:

98.5, 85.5, 80.0, 97, 93, 92.5, 93, 87, 88, 82

What is the range of her scores?
A) 17.0
B) 18.0
C) 18.5
D) 89.65

This is a question on calculating range. To calculate range, the lowest value in the data set is deducted from the highest value in the data set.

32) The median and mean of 9 numbers are 8 and 9 respectively. The 9 numbers are positive integers greater than zero. If each of the 9 numbers is increased by 2, which of the following must be true of the increased numbers?
A) The mean will be greater than before, but the median will remain the same.
B) The median will be greater than before, but the mean will remain the same.
C) Both the median and mean will be greater than before.
D) The median and mean will be the same as before, but the range will increase.

This is a question on interpreting distributions. Distribution is a measurement of how spread out the data is. Remember that for these types of questions, the quantity of items in the data set will usually not change. The question will normally state that all of the values in the set are going to increase or decrease by a certain amount. If all of the values in a data set are positive integers greater than zero and all of the values increase, the mean and median will also increase, but the range will not change. Conversely, if all of the values in such a data set decrease, the mean and median will also decrease, but the range will not change.

33) An owner of a carnival attraction draws teddy bears out of a bag at random to give to prize winners. She has 10 brown teddy bears, 8 white teddy bears, 4 black teddy bears, and 2 pink teddy bears when she opens the attraction at the start of the day. The first prize winner of the day receives a brown teddy bear. What is the probability that the second prize winner will receive a pink teddy bear?

A) $1/24$

B) $1/23$

C) $2/24$

D) $2/23$

This is a question on calculating basic probability. First of all, calculate how many items there are in total in the data set, which is also called the "sample space" or (S). Then reduce the data set if further items are removed. Probability can be expressed as a fraction. The number of items available in the total data set at the time of the draw goes in the denominator. The chance of the desired outcome, which is also referred to as the event or (E), goes in the numerator of the fraction. You can determine the chance of the event by calculating how many items are available in the subset of the desired outcome.

34) A magician pulls colored scarves out of a hat at random. The hat contains 5 red scarves and 6 blue scarves. The other scarves in the hat are green. If a scarf is pulled out of the hat at random, the probability that the scarf is red is $1/3$. How many green scarves are in the hat?
A) 3
B) 4
C) 5
D) 6

This question is asking you to determine the value missing from a sample space when calculating basic probability. This is like other problems on basic probability, but we need to work backwards to find the missing value. First, set up an equation to find the total items in the sample space. Then subtract the quantities of the known subsets from the total in order to determine the missing value.

35) The school board wants to poll a sample of students to get their opinions on dropping the music program in favor of having more sports programs. Which one of the following methods will result in the most statistically valid information about the opinions of all of the students at the high school?
A) To select ten students at random from each grade at the school
B) To speak to all of the members of the high school football team
C) To ask two members of each grade at random as they leave band practice
D) To give questionnaires out to the freshmen and sophomore students

This question is asking you about how best to use random sampling to draw conclusions about data. For information to be statistically valid, the data must be taken at random from a sample set of respondents that best represents the entire group.

36) The pictograph below shows the number of pizzas sold in one day at a local pizzeria. Cheese pizzas sold for $10 each, pepperoni pizzas sold for $12, and the total sales of all three types of pizza was $310. What is the sales price of one vegetable pizza?

Cheese	▼ ▼ ▼
Pepperoni	▼ ▼
Vegetable	▼

Each ▼ represents 5 pizzas.

A) $5
B) $8
C) $9
D) $12

This is an example of an exam question on interpreting data from pictographs. Each symbol on the pictograph represents a certain quantity of items, so remember to multiply by that amount in order to determine the totals for each group.

Problem solving, reasoning, and mathematical communication:

37) What number is next in this sequence? 2, 4, 8, 16
   A) 18
   B) 20
   C) 24
   D) 32

Questions on sequences like this one are problem solving and reasoning questions. Look at the possible relationships between the first two items in the series. Here, we can get to 4 from two by adding 2 or by doubling 2. Then try to perform these operations on the subsequent numbers to see what pattern emerges.

38) Use the information provided in the box below to answer the question that follows.

   - The police station is 10 miles away from the fire station
   - The fire station is 6 miles away from the hospital.

   Based on the information in the box, what conclusions can be made?
   A) The police station is no more than 6 miles away from the hospital.
   B) The police station is no more than 10 miles away from the hospital.
   C) The police station is exactly 6 miles away from the hospital.
   D) The police station is no more than 16 miles away from the hospital.

This is a mathematical communication problem involving deductive reasoning. Read the facts carefully and then make conclusions based on the information provided. You may find it helpful to draw diagrams to help you answer these types of questions.

39) A census shows that 1,008,942 people live in New Town, and 709,002 people live in Old Town. Which of the following numbers is the best estimate of how many more people live in New Town than in Old Town?
   A) 330,000
   B) 300,000
   C) 33,000
   D) 30,000

40) What is the best estimate for 1,198 ÷ 29 ?
   A) 37
   B) 40
   C) 60
   D) 400

> The two above questions are estimation and measurement problems involving estimating the results without doing full computations. To solve, we need to find estimates for the amounts in each question. Then subtract or divide using these estimates.

41) Use the information in the box below to answer the question that follows.

> - An orchard grows apples for resale. If the apple is 8 inches or more around, it is classified as grade A and sold to exclusive retailers.
> - If the apple measures less than 8 inches around, but more than 4 inches around, it is classified as grade B and sold to wholesalers.
> - If the apple measures 4 inches or less around it is classified as grade C.
> - Apples with a grade C classification are rejected for human consumption and are sold to animal food manufacturers.

If an apple measures exactly 4 inches around, which of the following statements could be true?
A) The apple will be classified as grade B.
B) The apple will be sold to exclusive retailers.
C) The apple will be sold to wholesalers.
D) The apple will not be eaten by people.

42) Use the information below to answer the question that follows.

> - Classes in the morning last for 45 minutes, but classes in the afternoon last for 50 minutes.
> - Lunch begins promptly at 12:30 pm and finishes promptly at 1:00 pm.
> - There are 3 classes after lunch and 4 classes before lunch.
> - There are no breaks between classes or between classes and lunch.

Which one of the following statements could be true?
A) Classes begin at 9:30am.
B) Classes begin at 10:00am.
C) The second class after lunch begins at 2:00pm.
D) The second class after lunch begins at 2:50pm.

> The two questions above are problem solving and reasoning questions. To answer these types of questions, read each of the answer choices one by one. Then compare each answer to the information provided to determine if that answer is true or false.

43) What is the largest possible product of two even integers whose sum is 22?
   A) 11
   B) 44
   C) 100
   D) 120

> You will see mathematical communication problems that ask you to perform multiplication or division on integers. Some of these problems may ask you to find an integer that meets certain mathematical requirements, like the problem above.

44) Which of the following methods of data representation would be best to represent the population of a certain city for each of the past ten years?
   A) scatterplot
   B) pie chart
   C) bar graph
   D) 3-D illustration

> For questions on data representation, remember the following:
> A scatterplot is used to see what kind of relationship exists between two items or events.
> A pie chart is useful to represent the amount or percentage of each group to the total of all groups.
> Bar graphs and line charts are usually used to represent changes to amounts over time.
> 3-D illustrations are useful for giving instructions on how to assemble something.

45) Which figure can be drawn by using the following algorithm?
   i.   Draw a line to the left 5 units long.
   ii. Turn the page right 90°.
   iii. Draw a line to the right 5 units long.
   iv. Turn the page right 90°.
   v.   Draw a line to the right 5 units long.

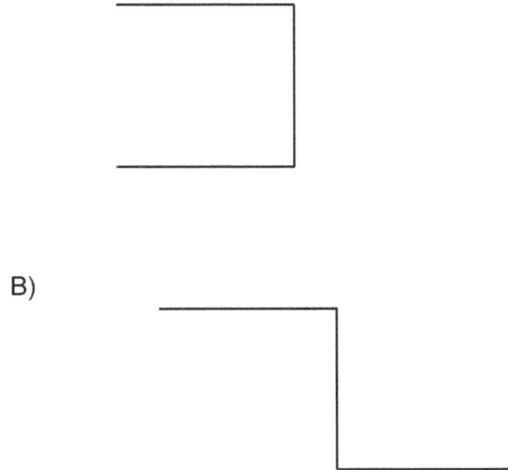

A)

B)

C)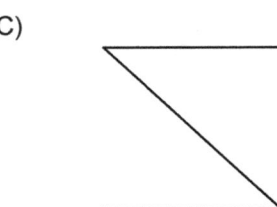

D)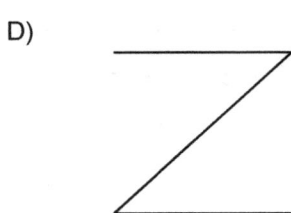

For questions on algorithms, draw the figure on a piece of scrap paper using the instructions provided. Be careful with the angle measurements. Although it may sound obvious, also be careful to distinguish left from right since this is more difficult to judge when you are turning.

Answer Key – NES Essential Academic Skills Practice Math Test 1

1) D
2) C
3) C
4) A
5) B
6) C
7) D
8) A
9) D
10) A
11) D
12) D
13) C
14) B
15) C
16) A
17) C
18) C
19) C
20) D
21) D
22) C
23) B

24) B
25) D
26) D
27) C
28) D
29) A
30) B
31) C
32) C
33) D
34) B
35) A
36) B
37) D
38) D
39) B
40) B
41) D
42) A
43) D
44) C
45) B

Solutions and Explanations to NES Essential Academic Skills Practice Math Test 1

1) The correct answer is D. Because two negatives make a positive, we know that $-(-5) = 5$. So, we can substitute this into the equation in order to solve it: $-(-5) + 3 = 5 + 3 = 8$

2) The correct answer is C. Multiply the numerators: $1 \times 2 = 2$. Then multiply the denominators: $3 \times 3 = 9$. These numbers form the new fraction: $2/9$

3) The correct answer is C.

STEP 1: Convert the first mixed number to an integer plus a fraction.

$$3\tfrac{1}{3} = 3 + \frac{1}{3}$$

STEP 2: Then multiply the integer by a fraction whose numerator and denominator are the same as the denominator of the existing fraction.

$$3 + \frac{1}{3} =$$

$$\left(3 \times \frac{3}{3}\right) + \frac{1}{3} =$$

$$\frac{9}{3} + \frac{1}{3}$$

STEP 3: Add the two fractions to get your new fraction.

$$\frac{9}{3} + \frac{1}{3} = \frac{10}{3}$$

Then repeat the steps to convert the second mixed number to a fraction, using the same steps that we have just completed for the first mixed number.

$$2\tfrac{1}{2} =$$

$$2 + \frac{1}{2} =$$

$$\left(2 \times \frac{2}{2}\right) + \frac{1}{2} =$$

$$\frac{4}{2} + \frac{1}{2} = \frac{5}{2}$$

Now that you have converted both mixed numbers to fractions, find the lowest common denominator and subtract to solve.

$$\frac{10}{3} - \frac{5}{2} =$$

$$\left(\frac{10}{3} \times \frac{2}{2}\right) - \left(\frac{5}{2} \times \frac{3}{3}\right) =$$

$$\frac{20}{6} - \frac{15}{6} =$$

$$\frac{5}{6}$$

4) The correct answer is A. There are no parentheses or exponents in this problem, so we need to direct our attention to the multiplication and division first. When you see a problem like this one, you need to do the multiplication and division from left to right. This means that you take the number to the left of the multiplication or division symbol and multiply or divide that number on the left by the number on the right of the symbol. So, in our problem we need to multiply –6 by 3 and then divide 4 by 2.

You can see the order of operations more clearly if you put in parentheses to group the numbers together.
–6 × 3 – 4 ÷ 2 =
(–6 × 3) – (4 ÷ 2) =
–18 – 2 = –20

5) The correct answer is B.

For this type of problem, do the operations inside the **parentheses** first.

$$\frac{5 \times (7-4)^2 + 3 \times 8}{5 - 6 \div (4-1)} =$$

$$\frac{5 \times (3)^2 + 3 \times 8}{5 - 6 \div 3}$$

Then do the operation on the **exponent**.

$$\frac{5 \times (3)^2 + 3 \times 8}{5 - 6 \div 3} =$$

$$\frac{5 \times (3 \times 3) + 3 \times 8}{5 - 6 \div 3}$$

$$\frac{5 \times 9 + 3 \times 8}{5 - 6 \div 3}$$

Then do the **multiplication** and **division**.

$$\frac{5 \times 9 + 3 \times 8}{5 - 6 \div 3} =$$

$$\frac{(5 \times 9) + (3 \times 8)}{5 - (6 \div 3)} =$$

$$\frac{45 + 24}{5 - 2}$$

Then do the **addition** and **subtraction**.

$$\frac{45+24}{5-2} = \frac{69}{3}$$

In this case, we can then simplify the fraction since both the numerator and denominator are divisible by 3.

$$\frac{69}{3} = 69 \div 3 = 23$$

6) The correct answer is C.

STEP 1: You can simplify the first fraction because both the numerator and denominator are divisible by 3: $3/6 \div 3/3 = 1/2$

STEP 2: Then divide the denominator of the second fraction ($x/14$) by the denominator of the simplified fraction ($1/2$) from above: $14 \div 2 = 7$

STEP 3: Now, multiply the number from step 2 by the numerator of the fraction we calculated in step 1 in order to get your result: $1 \times 7 = 7$

You can check your answer as follows:
$3/6 = 7/14$
$3/6 \div 3/3 = 1/2$
$7/14 \div 7/7 = 1/2$

7) The correct answer is D. This problem is asking for the ratio of non-faulty mp3 players to the quantity of faulty mp3 players. Therefore, you must put the quantity of non-faulty mp3 players before the colon in the ratio. In this problem, 1% of the players are faulty.

1% × 100 = 1 faulty player in every 100 players
100 − 1 = 99 non-faulty players

So, the ratio is 99:1. As explained in the study tip after the question, the number before the colon and the number after the colon can be added together to get the total quantity.

8) The correct answer is A. The base number in this example is 11. So, we add the exponents:
$11^5 \times 11^3 = 11^{(5+3)} = 11^8$

9) The correct answer is D. Remember that the decimal number can have only one digit and the other numbers must be decimals. So $784 = 7.84 \times 10 \times 10 = 7.84 \times 10^2$

10) The correct answer is A.

$4x^2 + 2xy - y^2 =$
$(4 \times 2^2) + (2 \times 2 \times -2) - (-2^2) =$
$(4 \times 2 \times 2) + (2 \times 2 \times -2) - (-2 \times -2) =$
$(4 \times 4) + (2 \times -4) - (4) =$
$16 + (-8) - 4 =$
$16 - 12 = 4$

11) The correct answer is D.

STEP 1: Factor the equation.
$x^2 + 6x + 8 = 0$
$(x + 2)(x + 4) = 0$

STEP 2: Now substitute 0 for x in the first pair of parentheses.
$(0 + 2)(x + 4) = 0$
$2(x + 4) = 0$
$2x + 8 = 0$
$2x + 8 - 8 = 0 - 8$
$2x = -8$
$2x \div 2 = -8 \div 2$
$x = -4$

STEP 3: Then substitute 0 for x in the second pair of parentheses.
$(x + 2)(x + 4) = 0$
$(x + 2)(0 + 4) = 0$
$(x + 2)4 = 0$
$4x + 8 = 0$
$4x + 8 - 8 = 0 - 8$
$4x = -8$
$4x \div 4 = -8 \div 4$
$x = -2$

12) The correct answer is D. For inequality problems like this, look to see if both of the equations have any variables or terms in common. In this problem, both equations contain $x - 2$. The cost of one movie ticket is represented by y, and y is equal to $x - 2$. Therefore, we can substitute values from one equation to another.

$x - 2 > 5$
$y > 5$

If two tickets are being purchased, we need to solve for $2y$.
$y \times 2 > 5 \times 2$
$2y > 10$

13) The correct answer is C. For some basic equation problems, you will see two equations which have the same two variables, like J and T in this problem. In order to solve the problem, take the second equation and isolate J on one side of the equation. By doing this, you define variable J in terms of variable T.
$J + 2T = \$40$
$J + 2T - 2T = \$40 - 2T$
$J = \$40 - 2T$

Now substitute $\$40 - 2T$ for variable J in the first equation to solve for variable T.
$2J + T = 50$
$2(40 - 2T) + T = 50$
$80 - 4T + T = 50$
$80 - 3T = 50$
$80 - 3T + 3T = 50 + 3T$
$80 = 50 + 3T$
$80 - 50 = 50 - 50 + 3T$

30 = 3T
30 ÷ 3 = 3T ÷ 3
10 = T

So, now that we know that a T-shirt costs $10, we can substitute this value in one of the equations in order to find the value for the jeans, which is variable J.
2J + T = 50
2J + 10 = 50
2J + 10 − 10 = 50 − 10
2J = 40
2J ÷ 2 = 40 ÷ 2
J = 20

Now solve for Sarah's purchase. If she purchased one pair of jeans and one T-shirt, then she paid: $10 + $20 = $30

14) The correct answer is B. To solve this type of problem, do multiplication on the items in parentheses first.

3x − 2(x + 5) = −8
3x − 2x − 10 = −8

Then deal with the integers by putting them on one side of the equation.
3x − 2x − 10 + 10 = −8 + 10
3x − 2x = 2

Then solve for x.
3x − 2x = 2
1x = 2
x = 2

15) The correct answer is C. The problem states the y intercept, so use the slope-intercept formula to solve.
y = mx + b
165 = m5 + 15
165 − 15 = m5 + 15 − 15
150 = m5
150 ÷ 5 = m5 ÷ 5
30 = m

16) The correct answer is A. Find the solutions for the x and y intercepts separately as shown below. First, substitute 0 for y in order to find the x intercept.

$x^2 + 4y^2 = 64$
$x^2 + (4 \times 0) = 64$
$x^2 + 0 = 64$
$x^2 = 64$
$x = 8$

Then substitute 0 for x in order to find the y intercept.
$x^2 + 4y^2 = 64$
$(0 \times 0) + 4y^2 = 64$

$0 + 4y^2 = 64$
$4y^2 \div 4 = 64 \div 4$
$y^2 = 16$
$y = 4$
So, the x intercept is (8, 0) and the y intercept is (0, 4).

17) The correct answer is C. As the quantity of sugar increases, the amount of sleep also increases. A positive linear relationship therefore exists between the two variables. This is represented in chart C since the amount of sleep is greater when the amount of sugar consumed is higher.

18) The correct answer is C. We can see that the line does not begin on exactly on (5, 5), nor does it begin on (5, 9) because the first point is slightly below the horizontal line for y = 5. Therefore, we can rule out answers A and D.

If we look at x = 20 on the graph, we can see that y = 18 at this point.
As stated in the study tip, we can express this as the function: $f(x) = x \times 0.9$

Putting in the values of x from chart (C), we get the following:
$5 \times 0.9 = 4.5$
$10 \times 0.9 = 9$
$15 \times 0.9 = 13.5$
$20 \times 0.9 = 18$

19) The correct answer is C. The sum of all three angles inside a triangle is always 180 degrees. So, we need to deduct the degrees given from 180° to find out the total degrees of the two other angles: 180° − 36° = 144°

Now divide this result by two in order to determine the degrees for each angle: 144° ÷ 2 = 72°

20) The correct answer is D. The area of a rectangle is equal to its length times its width. This football field is 30 yards wide and 100 yards long, so now we can substitute the values.
rectangle area = width × length
rectangle area = 30 × 100
rectangle area = 3000

21) The correct answer is D. Substitute the values into the formula in order to find the solution for this problem:

$\sqrt{A^2 + B^2}$ = C

$\sqrt{5^2 + 12^2}$ = C

$\sqrt{25 + 144}$ = C

$\sqrt{169}$ = C

13 cm = C

22) The correct answer is C. The base length of the triangle described in the problem, which is line segment YZ, is not given. So, we need to calculate the base length using the Pythagorean Theorem. According to the Pythagorean Theorem, the length of the hypotenuse is equal to the square root of the sum of the squares of the two other sides.

$$\sqrt{4^2 + base^2} = 5$$
$$\sqrt{16 + base^2} = 5$$

Now square each side of the equation in order to solve for the base length.

$$\sqrt{16 + base^2} = 5$$
$$(\sqrt{16 + base^2})^2 = 5^2$$
$$16 + base^2 = 25$$
$$16 - 16 + base^2 = 25 - 16$$
$$base^2 = 9$$
$$\sqrt{base^2} = \sqrt{9}$$
$$base = 3$$

Now solve for the area of the triangle.
triangle area = (base × height) ÷ 2
triangle area = (3 × 4) ÷ 2
triangle area = 12 ÷ 2
triangle area = 6

23) The correct answer is B. Write out the formula: (length × 2) + (width × 2). Then substitute the values: (5 × 2) + (3 × 2) = 10 + 6 = 16

24) The correct answer is B. Substitute the value of the diameter into the formula to calculate the circumference.
circumference = diameter × π
circumference = 12π

25) The correct answer is D. To calculate the volume of a box, you need the formula for a rectangular solid: volume = base × width × height

Now substitute the values from the problem into the formula.
volume = 20 × 15 × 25
volume = 7500

26) The correct answer is D. You will need to understand the basic concepts of United States' measurements for the exam. Remember that large fluid items are measured in pints and quarts, while large solid items are measured in pounds, or tons in the case of extremely heavy quantities. Feet and inches are linear measurements; they are not used for weight. Liters and gallons are measures of liquid substances. Horsepower measures the strength of an engine. Tons measure the weight of heavy items, so tons would be suitable for measuring the weight of a car. Note that one ton is equal to two thousand pounds.

27) The correct answer is C. 1 yard = 36 inches, so we need to multiply by the number of yards:
6 yards × 36 inches = 216 inches

28) The correct answer is D. Set up your equation to calculate the average, using x for the age of the 5th sibling:

$(2 + 5 + 7 + 12 + x) \div 5 = 8$
$(2 + 5 + 7 + 12 + x) \div 5 \times 5 = 8 \times 5$
$(2 + 5 + 7 + 12 + x) = 40$
$26 + x = 40$
$26 - 26 + x = 40 - 26$
$x = 14$

29) The correct answer is A. The mode is the number that occurs the most frequently in the set. Our data set is: 1, 1, 3, 2, 4, 3, 1, 2, 1. The number 1 occurs 4 times in the set, which is more frequently than any other number in the set, so the mode is 1.

30) The correct answer is B. The problem provides the number set: 8.19, 7.59, 8.25, 7.35, 9.10
First of all, put the numbers in ascending order: 7.35, 7.59, 8.19, 8.25, 9.10
Then find the one that is in the middle: 7.35, 7.59, **8.19**, 8.25, 9.10

31) The correct answer is C. To calculate the range, the low number in the set is deducted from the high number in the set. The problem set is: 98.5, 85.5, 80.0, 97, 93, 92.5, 93, 87, 88, 82. The high number is 98.5 and the low number is 80, so the range is 18.5 (98.5 − 80 = 18.5)

32) The correct answer is C. If each number in the set is increased by 2, the mean will increase by 2. Here we have 9 numbers in the set, so the overall increase in the total of the values (2 × 9 = 18) will be divided equally among all nine items in the set (18 ÷ 9 = 2) when the mean is calculated. Since each of the numbers increases by 2, the median number will also increase by 2. So, both the median and mean will be greater than before.

33) The correct answer is D. You need to determine the amount of possible outcomes at the start of the day first of all. The owner has 10 brown teddy bears, 8 white teddy bears, 4 black teddy bears, and 2 pink teddy bears when she opens the attraction at the start of the day. So, at the start of the day, she has 24 teddy bears: 10 + 8 + 4 + 2 = 24. Then you need to reduce this amount by the quantity of items that have been removed. The problem tells us that she has given out a brown teddy bear, so there are 23 teddy bears left in the sample space: 24 − 1 = 23
The event is the chance of the selection of a pink teddy bear. We know that there are two pink teddy bears left after the first prize winner receives his or her prize. Finally, we need to put the event (the number representing the chance of the desired outcome) in the numerator and the number of possible remaining combinations (the sample space) in the denominator.
So the answer is $^2/_{23}$.

34) The correct answer is B. First, we will use variable T as the total number of items in the set. The probability of getting a red scarf is $^1/_3$. So, set up an equation to find the total items in the data set:

$$\frac{5}{T} = \frac{1}{3}$$

$$\frac{5}{T} \times 3 = \frac{1}{3} \times 3$$

$$\frac{5}{T} \times 3 = 1$$

$$\frac{15}{T} = 1$$

$$\frac{15}{T} \times T = 1 \times T$$

$$15 = T$$

We have 5 red scarves, 6 blue scarves, and x green scarves in the data set that make up the total sample space, so now subtract the amount of red and blue scarves from the total in order to determine the number of green scarves.
$5 + 6 + x = 15$

$11 + x = 15$

$11 - 11 + x = 15 - 11$

$x = 4$

35) The correct answer is A. Our entire group in this problem is all of the students at the high school. So, it would be best to select ten students at random from each grade at the school. The other answer choices would be biased in favor of members of certain groups, namely football players (answer B), band participants (answer C), and younger students (answer D).

36) The correct answer is B. First, determine how many cheese and pepperoni pizzas were sold. Each triangle symbol represents 5 pizzas. Therefore, 15 cheese pizzas were sold: 3 symbols on the pictograph × 5 pizzas per symbol = 15 cheese pizzas. We also know that 10 pepperoni pizzas were sold: 2 symbols on the pictograph × 5 pizzas per symbol = 10 pepperoni pizzas. Then determine the sales value of these two types of pizzas based on the prices stated in the problem:
(15 cheese pizzas × $10 each) + (10 pepperoni pizzas × $12 each) = $150 + $120 = $270

The remaining amount is allocable to the vegetable pizzas:
Total sales of $310 − $270 = $40 worth of vegetable pizzas

Since each triangle represents 5 pizzas, 5 vegetable pizzas were sold. We calculate the price of the vegetable pizzas as follows:
$40 worth of vegetable pizzas ÷ 5 vegetable pizzas sold = $8 per vegetable pizza

37) The correct answer is D. Try to find the pattern of relationship between the numbers.
Here, we can see this pattern:
2 × 2 = 4
4 × 2 = 8
8 × 2 = 16
In other words, the next number in the sequence is always double the previous number.
Therefore the answer is: 16 × 2 = 32

38) The correct answer is D. For questions about distance like this one, keep in mind that the locations may or may not lie on a straight line.

For example, the locations could be laid out like this:

Police station 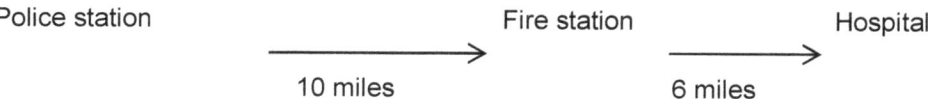 Fire station Hospital

10 miles   6 miles

In the layout above, the police station would be 16 miles from the hospital.

However, the locations could also be laid out like this:

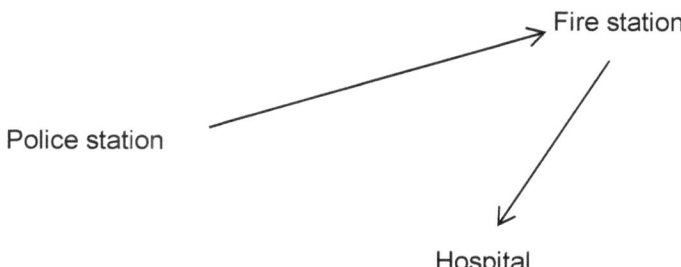

Fire station

Police station

Hospital

We can see that the locations will be the farthest from each other if they are laid out on a straight line as in the first example above.

In other words, a person could always go to the hospital by traveling to the fire station from the police station (10 miles) and then traveling from the fire station to the hospital (6 miles).

Therefore, the police station would never be more than 16 miles away from the hospital, regardless of the layout.

39) The correct answer is B. This is another type of estimation question. The problem tells us that 1,008,942 people live in New Town, and 709,002 people live in Old Town.

STEP 1: We need to round the numbers up or down to the nearest thousand as needed. So 1,008,942 is rounded to 1,009,000 and 709,002 is rounded to 709,000.

STEP 2: Then subtract the second figure from the first figure in order to get your result. 1,009,000 − 709,000 = 300,000

40) The correct answer is B. We can see from the answer choices that we are rounding to the nearest increment of 10. So 1,198 is rounded up to 1,200 and 29 is rounded up to 30. Now do the operation: 1,200 ÷ 30 = 40

41) The correct answer is D. Read the facts of problems like this one very carefully. The facts provided in the problem tell us that if an apple measures 4 inches or less around it is classified as grade C, which is sold to animal food manufacturers. The apple in this problem is exactly 4 inches, so it is a grade C apple. Therefore, the apple will not be eaten by people, but by animals.

42) The correct answer is A. If classes last for 45 minutes and there are 4 classes before lunch, the morning classes last for 3 hours in total. If lunch is at 12:30, it is therefore possible for classes to begin at 9:30.

43) The correct answer is D. For problems that ask you to find the largest possible product of two even integers, first you need to divide the sum by 2. The sum in this problem is 22, so divide by 2.

22 ÷ 2 = 11

Now take the result from this division and find the 2 nearest even integers that are 1 number higher and lower.

11 + 1 = 12

11 − 1 = 10

Then multiply these two numbers together in order to get the product: 12 × 10 = 120

44) The correct answer is C. In the bar graph, you will have bars for all of the years, with each bar representing the amount of the population that year. This would make it easy to see the changes to the population over time. A line graph would also be suitable, although that is not a choice for this question.

45) The correct answer is B.

i. Draw a line to the left 5 units long.

ii. Turn the page right 90°.

iii. Draw a line to the right 5 units long.

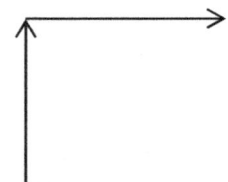

iv. Turn the page right 90°.

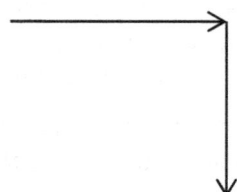

v. Draw a line to the right 5 units long.

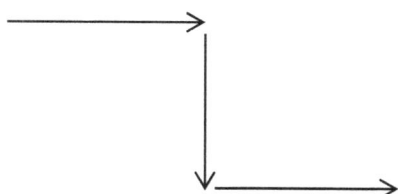

NES Essential Academic Skills Practice Math Test 2 with Study Tips

Number properties and number operations:

1) $\dfrac{1}{5} \div \dfrac{4}{7} = ?$

   A) $\dfrac{4}{20}$

   B) $\dfrac{7}{20}$

   C) $\dfrac{4}{35}$

   D) $\dfrac{5}{35}$

> You will also need to know how to divide fractions for the exam. To divide fractions, invert the second fraction by putting the denominator on the top and numerator on the bottom. Then multiply the fractions, as demonstrated in practice test 1.

2) What is $\dfrac{1}{9} + \dfrac{9}{27}$?

   A) $\dfrac{12}{27}$

   B) $\dfrac{9}{27}$

   C) $\dfrac{3}{27}$

   D) $\dfrac{10}{36}$

> In some fraction problems, you will have to find the lowest common denominator. In other words, before you add or subtract fractions, you have to change them so that the bottom numbers in each fraction are the same. You do this by multiplying the numerator by the same number that you used when multiplying to get the new denominator for the fraction.

3) Simplify: $\dfrac{12}{27}$

   A) $\dfrac{1}{3}$

   B) $\dfrac{3}{4}$

C) $\frac{3}{9}$

D) $\frac{4}{9}$

> You will also need to know how to simplify fractions for your exam. To simplify fractions, look to see what factors are common to both the numerator and denominator. Factoring is like taking a number apart. So, what numbers can we multiply together to get 12? What numbers can we multiply together to get 27?

4) Consider a class which has n students. In this class, $t\%$ of the students subscribe to digital TV packages. Which of the following equations represents the number of students who do not subscribe to any digital TV package?
   A) $100(n - t)$
   B) $(100\% - t\%) \times n$
   C) $(100\% - t\%) \div n$
   D) $(1 - t)n$

> You will have to calculate percentages and decimals on the exam, as well as use percentages and decimals to solve other types of math problems or to create equations. Percentages can be expressed by using the symbol %. They can also be expressed as fractions or decimals.

5) A company purchases cell phones at a cost of x and sells the cell phones at four times the cost. Which of the following represents the profit made on each cell phone?
   A) x
   B) $3x$
   C) $4x$
   D) $3 - x$

> You will see problems on the test that ask you to set up mathematical equations from basic information. To set up an equation, read the problem carefully and then express the facts in terms of a mathematical equation. The problem tells us that cell phones sell for four times the cost, so "four times" means that we have to multiply. For this problem, profit is calculated by taking the sales price and subtracting the cost.

6) Which of the following shows the numbers ordered from least to greatest?
   A) $-1/4, 1/8, 1/6, 1$
   B) $-1/4, 1/8, 1, 1/6$
   C) $-1/4, 1/6, 1/8, 1$
   D) $-1/4, 1, 1/8, 1/6$

> In order to answer questions on ordering fractions and other numbers from least to greatest or greatest to least, remember these principles: (a) Negative numbers are less than positive numbers; (b) When two fractions have the same numerator, the fraction with the smaller number in the denominator is the larger fraction.

7) The temperature on Saturday was 62° F at 5:00 PM and 38° F at 11:00 PM. If the temperature fell at a constant rate on Saturday, what was the temperature at 9:00 PM?
    A) 58° F
    B) 54° F
    C) 50° F
    D) 46° F

> This question assesses your knowledge of performing operations on integers. Here, we have to perform the operations of subtraction, multiplication, and division.

8) The Smith family is having lunch in a diner. They buy hot dogs and hamburgers to eat. The hot dogs cost $2.50 each, and the hamburgers cost $4 each. They buy 3 hamburgers. They also buy hot dogs. The total value of their purchase is $22. How many hot dogs did they buy?
    A) 2
    B) 3
    C) 4
    D) 5

> This question assesses your knowledge of performing addition, subtraction, multiplication, and division on integers in a single practical problem. Set up an equation, with the number of hot dogs represented by *D* and the number of hamburgers represented by *H*.

9) $10^6 \div 10^4 = ?$
    A) 10^{24}
    B) 10^2
    C) 20^{24}
    D) 20^2

> Remember to subtract the exponents when you divide the base numbers.
>
> Although not needed for this problem, you will also need to know the following exponent properties for the exam.
>
> **Zero exponent:** Any number to the power of zero is equal to 1.
>
> $$\text{Example: } 9^0 = 1$$
>
> **Negative exponents:** Remove the negative sign on the exponent by expressing the number as a fraction, with 1 as the numerator. Then place the number with the exponent in the denominator.
>
> $$\text{Example: } x^{-2} = \frac{1}{x^2}$$
>
> **Fractional exponents:** Place the base number inside the radical sign. The denominator of the exponent is the nth root of the radical. The numerator is new exponent.
>
> $$\text{Example: } x^{3/7} = (\sqrt[7]{x})^3$$

Algebra and graphing:

10) Solve for x: $\frac{3}{4}x - 2 = 4$

   A) $\frac{8}{3}$

   B) $\frac{1}{8}$

   C) 8

   D) −8

> This is a problem requiring you to solve an expression that contains a single variable, a fraction, and integers. First, isolate the integers and then eliminate the fraction. Finally, divide to find the value of the variable.

11) If $2(3x - 1) = 4(x + 1) - 3$, what is the value of x?

   A) $x = 3/2$

   B) $x = 2/3$

   C) $x = 4/3$

   D) $x = 3/4$

> This problem requires you to solve an algebraic expression that contains one variable (x) on both sides of the equation. When the variable is used on both sides of the equation, you should perform the multiplication on the parentheticals first. Then isolate x to solve the problem.

12) Consider the inequality: $-3x + 14 < 5$
   Which of the following values of x is a possible solution to the inequality above?

   A) −3.1

   B) 2.25

   C) 2.65

   D) 4.35

> This question is assessing your understanding of inequalities. When dealing with inequalities, we first need to place the integers on one side of the inequality. Then deal with any negative numbers. Remember that when you divide or multiply by a negative number in inequality problems, you need to reverse the way that the inequality sign points.

13) Which of the following equations is equivalent to $\frac{x}{5} + \frac{y}{2}$?

   A) $\frac{x+y}{7}$

   B) $\frac{2x+5y}{10}$

   C) $\frac{5x+2y}{10}$

   D) $\frac{5y}{2x}$

This problem is asking you to find an equivalent expression for a mathematical equation that contains fractions. To add fractions, find the lowest common denominator first and then add the numerators.

14) If $W = \dfrac{XY}{Z}$, then $Z = ?$

    A) $\dfrac{XY}{W}$

    B) $\dfrac{W}{XY}$

    C) $\dfrac{1}{XY}$

    D) $\dfrac{Y}{XW}$

To find this equivalent expression, isolate the variable that the question is asking for, performing multiplication and division as necessary.

15) An internet provider sells internet packages based on monthly rates. The price for the internet service depends on the speed of the internet connection. The chart that follows indicates the prices of the various internet packages.

Price in dollars (P)	10	20	30	40
Gigabyte speed (s)	2	4	6	8

Which equation represents the prices of these internet packages?
A) $P = (s - 5) \times 5$
B) $P = (s + 5) \times 5$
C) $P = 5 \div s$
D) $P = s \times 5$

To set up a basic equation, remember to read the problem carefully and then express the facts in terms of an algebraic equation. For the question above, divide the dollars by the speed to find the mathematical relationship.

16) $(3x - 2y)^2 = ?$
A) $9x^2 + 4y^2$
B) $9x^2 - 6xy^2 + 4y^2$
C) $9x^2 - 12xy + 4y^2$
D) $9x^2 + 12xy + 4y^2$

> Multiplying Polynomials Using the FOIL Method – Polynomials are algebraic expressions that contain integers, variables, and variables which are raised to whole-number positive exponents. You will also see binomial problems on the test. Binomial multiplication problems will frequently be in this format: (a +b)(c + d).
> Multiply the terms in the parentheses in this order: **First – Outside – Inside – Last**
> So, (a +b)(c + d) = (a × c) + (a × d) + (b × c) + (b × d)

17) How many solutions exist for the following equation? $x^2 + 8 = 0$
    A) 0
    B) 1
    C) 2
    D) 4

> You will see questions on the exam that give you an equation and then ask you how many solutions there are for the equation provided. You will need to consider both positive and negative numbers as potential solutions.

18) Consider two stores in a town. The first store is a grocery store. The second is a pizza place where customers collect their pizzas after they order them online. The grocery store is represented by the coordinates (−4, 2) and the pizza place is represented by the coordinates (2,−4). If the grocery store and the pizza place are connected by a line segment, what is the midpoint of this line?
    A) (1, 1)
    B) (−1, −1)
    C) (2, 2)
    D) (−2, −2)

> You may be asked to calculate the midpoint of two points on a graph. Remember that you divide the sum of the two points by 2 because the midpoint is the halfway mark between the two points on the line. The two points are represented by the coordinates (x_1, y_1) and (x_2, y_2). The midpoints of two points on a two-dimensional graph are calculated by using the midpoint formula:
>
> $$(x_1 + x_2) \div 2 , (y_1 + y_2) \div 2$$

Geometry and measurement:

19) In the figure below, x and y are parallel lines, and line z is a transversal crossing both x and y. Which three angles are equal in measure? You may select more than one answer.

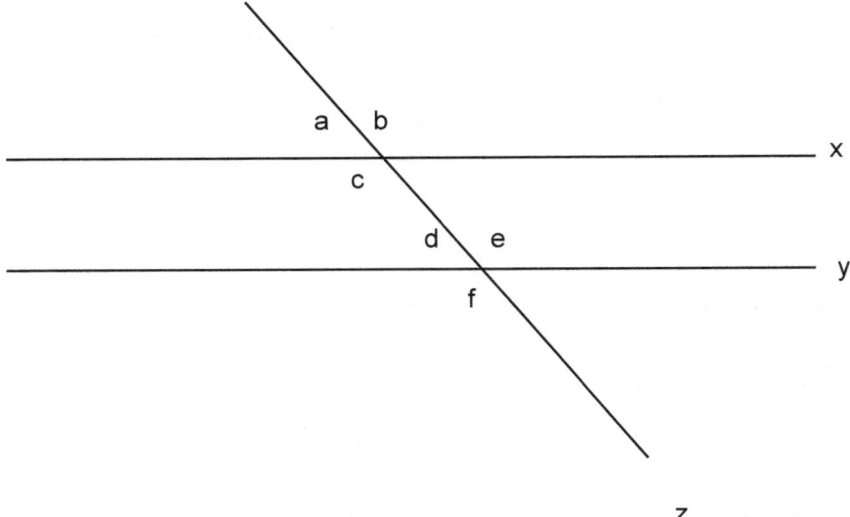

A) ∠a, ∠b, ∠c
B) ∠b, ∠c, ∠f
C) ∠a, ∠d, ∠e
D) ∠a, ∠d, ∠f

When two parallel lines are cut by a transversal (a straight line that runs through both of the parallel lines), 4 pairs of opposite (non-adjacent) angles are formed and 4 pairs of corresponding angles are formed. The opposite angles will be equal in measure, and the corresponding angles will also be equal in measure.

20)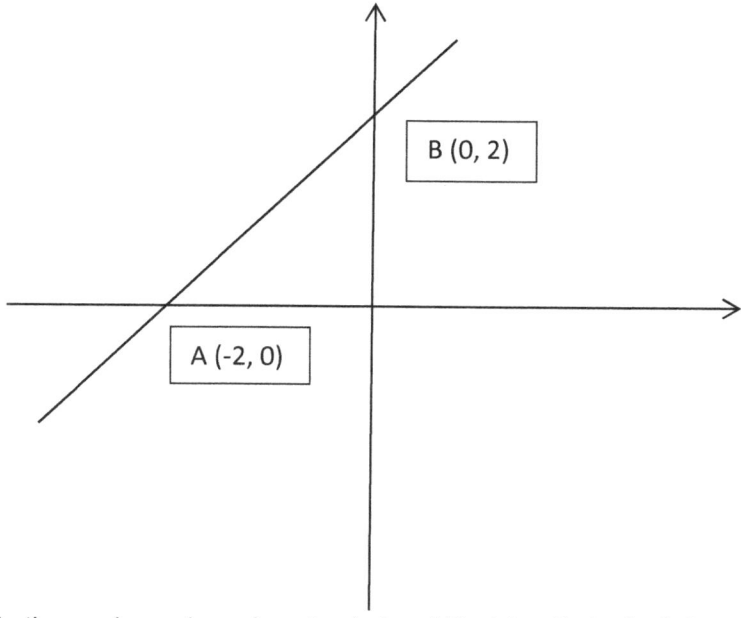

The line in the *xy* plane above is going to be shifted 5 units to the left and 4 units up. What are the coordinates of point B after the shift?
A) (−5, 6)
B) (5, 6)
C) (5, 4)
D) (−7, 4)

This question involves transposing the coordinates of a linear equation.

Remember these rules on transpositions:

x coordinate moved to the left – deduct the units from the original *x* coordinate

x coordinate moved to the right – add the units to the original *x* coordinate

y coordinate moved down – deduct the units from the original *y* coordinate

y coordinate moved up – add the units to the original *y* coordinate

---

21) Pat wants to put wooden trim around the floor of her family room. Each piece of wood is 1 foot in length. The room is rectangular and is 12 feet long and 10 feet wide. How many pieces of wood does Pat need for the entire perimeter of the room?
A) 22
B) 44
C) 100
D) 120

This is an advanced question on calculating perimeter. First, determine the perimeter using the formula: (2 × width) + (2 × height) = perimeter. Then divide to find the number of pieces needed to cover the perimeter.

22) Find the volume of a cone which has a radius of 3 and a height of 4.
A) 4π
B) 12π
C) $4π/3$
D) $3π/4$

To determine the volume of a cone, use the following formula:
Cone volume = (π × radius² × height) ÷ 3

23) The diagram below depicts a cell phone tower. The height of the tower from point B at the center of its base to point T at the top is 30 meters, and the distance from point B of the tower to point A on the ground is 18 meters. What is the approximate distance from point A on the ground to the top (T) of the cell phone tower?

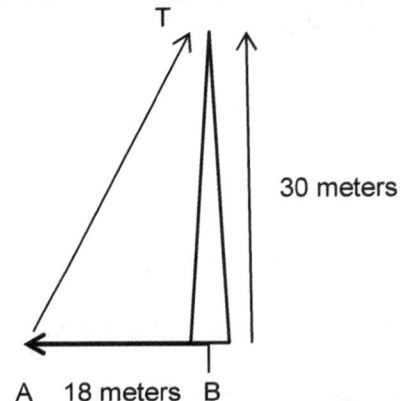

A) 10 meters
B) 20 meters
C) 30 meters
D) 35 meters

We need to use the Pythagorean Theorem to solve the problem. The Pythagorean Theorem deals with right triangles. The theorem helps us to calculate the length of the hypotenuse, which is the side opposite the right angle (The right angle is at the 90° corner of the triangle.) The hypotenuse is called side C in the formula for the Pythagorean Theorem. The theorem states that the length of the hypotenuse is equal to the square root of the sum of the squares of the lengths of the two other sides (*A* and *B*). Use the following formula to calculate the length of the hypotenuse:

$$\sqrt{A^2 + B^2} = C$$

24) Which of the following statements about isosceles triangles is true?
   A) Isosceles triangles have two equal sides.
   B) When an altitude is drawn in an isosceles triangle, two equilateral triangles are formed.
   C) The base of an isosceles triangle must be shorter than the length of each of the other two sides.
   D) The sum of the measurements of the interior angles of an isosceles triangle must be equal to 360°.

This question assesses your knowledge of the rules for triangles and angles. Remember these principles on angles and triangles for your exam:

The sum of all three angles in any triangle must be equal to 180 degrees.

An isosceles triangle has two equal sides and two equal angles.

An equilateral triangle has three equal sides and three equal angles.

Angles that have the same measurement in degrees are called congruent angles.

Equilateral triangles are sometimes called congruent triangles.

Two angles are supplementary if they add up to 180 degrees. This means that when the two angles are placed together, they will form a straight line on one side.

Two angles are complementary (sometimes called adjacent angles) if they add up to 90 degrees. This means that the two angles will form a right triangle.

A parallelogram is a four-sided figure in which opposite sides are parallel and equal in length. Each angle will have the same measurement as the angle opposite to it, so a parallelogram has two pairs of opposite angles.

The sides of a 30° - 60° - 90° triangle are in the ratio of 1: $\sqrt{3}$: 2.

---

25) In the figure below, XY and WZ are parallel, and lengths are provided in units. What is the area of trapezoid WXYZ in square units?

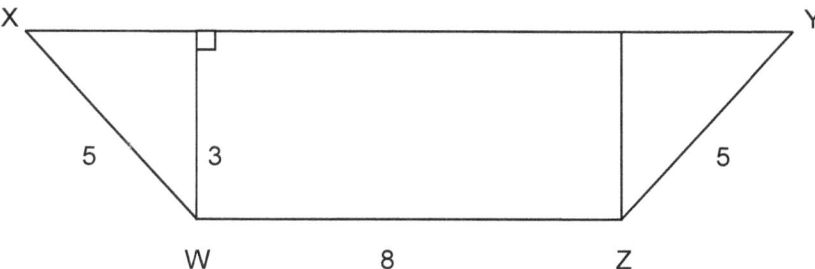

A) 24
B) 30
C) 34
D) 36

Now we are going to look at advanced questions on hybrid shapes.
First, calculate the area of the central rectangle.
Remember that the area of a rectangle is length times height.
Then calculate the area of each of the triangles on each side of the central rectangle.
Remember that the area of a triangle is base times height divided by 2.
The total area is the area of the main rectangle plus the area of each of the two triangles.

26) Which of the following dimensions would be needed in order to find the area of the figure?

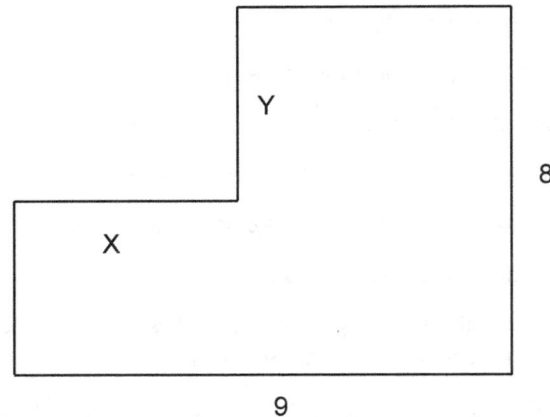

A) X only
B) Y only
C) Both X and Y
D) Either X or Y

This question is also asking you to calculate the area of a hybrid shape. To solve problems like this one, try to visualize two rectangles. The first rectangle would measure 8 × 9 and the second rectangle would measure X × Y.

27) In the figure below, the lengths of KL, LM, and KN are provided in units. What is the area of triangle NLM in square units?

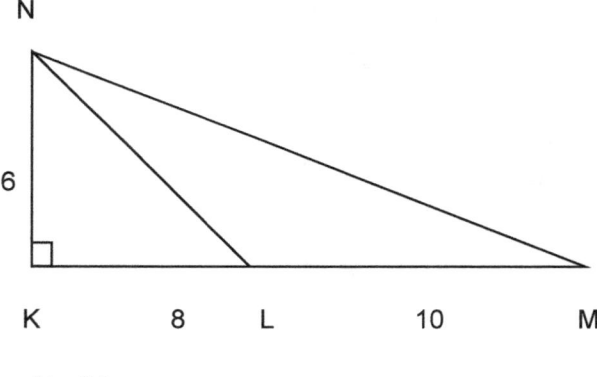

A) 24
B) 30

42

C) 48
D) 54

> Calculate the area of triangle NKM. Then, calculate the area of the area of triangle NKL. The remaining triangle NLM is then calculated by subtracting the area of triangle NKL from triangle NKM.

Probability and statistics:

28) Becky rolls a fair pair of six-sided dice. One of the die is black and the other is red. Each die has values from 1 to 6. What is the probability that Becky will roll a 4 on the red die and a 5 on the black die?

   A) $1/36$

   B) $2/36$

   C) $1/12$

   D) $2/12$

> This is an advanced problem on understanding probability models. For these questions, you will usually have two items, like two dice or a coin and a die. Each item will have various outcomes, like heads or tails for the coin or the different numbers on the die. To solve problems like this one, it is usually best to write out the possible outcomes in a list. This will help you visualize the number of possible outcomes that make up the sample space. Then circle or highlight the events from the list to get your answer.

29) 110 students took a math test. The mean score for the 60 female students was 95, while the mean score for the 50 male students was 90. Which figure below best approximates the mean test score for all 110 students in the class?
   A) 55
   B) 90
   C) 92.5
   D) 92.73

> For advanced questions on averages like this one, perform these operations:
> Find the total points for all the females and the total points for all the males.
> Then add these two amounts together and divide by the total number of students in the class.

30) Return on investment (ROI) percentages are provided for seven companies. The ROI will be negative if the company operated at a loss, but the ROI will be a positive value if the company operated at a profit. The ROI's for the seven companies were: –2%, 5%, 7.5%, 14%, 17%, 1.3%, –3%. Which figure below best approximates the mean ROI for the seven companies?
   A) 2%
   B) 5.7%

C) 6.25%
D) 7.5%

31) A group of families had the following household incomes on their tax returns: $65000, $52000, $125000, $89000, $36000, $84000, $31000, $135000, $74000, and $87000. What is the range?
A) 74000
B) 77800
C) 79000
D) 104000

32) The graph below shows the relationship between the total number of chicken sandwiches a restaurant sells and the total sales in dollars for the chicken sandwiches. What is the sales price per chicken sandwich?

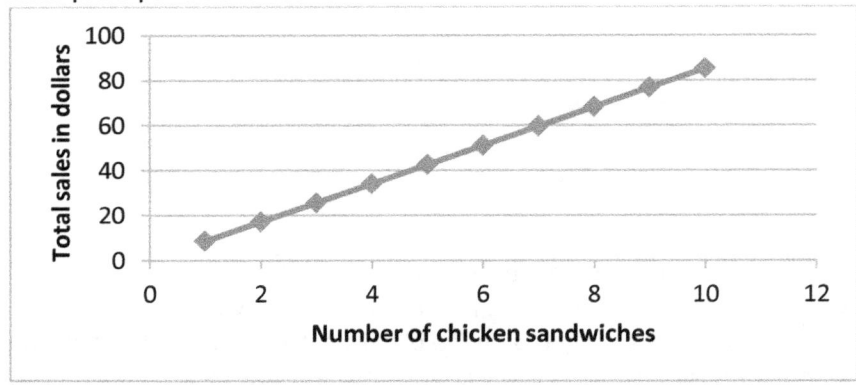

A) $4.00
B) $8.00
C) $8.50
D) $9.50

This is an example of a question that asks you to interpret a line graph in order to determine the price per unit of an item. To solve the problem, look at a specific point on the graph and then divide the total sales in dollars by the total quantity sold in order to get the price per unit.

33) In Brown County Elementary School, parents are advised to have their children vaccinated against five childhood diseases. According to the chart below, how many children were vaccinated against at least three diseases?

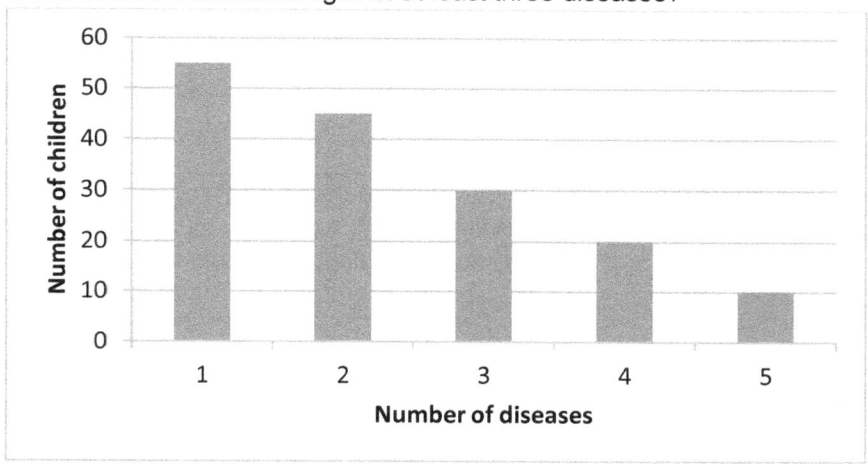

A) 30
B) 50
C) 60
D) 100

> For questions that ask you to interpret bar graphs or histograms, you need to read the problem carefully to determine what is represented on the horizontal axis (bottom) and the vertical axis (left side) of the graph. Histograms are like bar graphs, except they represent groups of data. We will look at histograms in the next practice test.

34) A zoo has reptiles, birds, quadrupeds, and fish. At the start of the year, they have a total of 1,500 creatures living in the zoo. The pie chart below shows percentages by category for the 1,500 creatures at the start of the year.

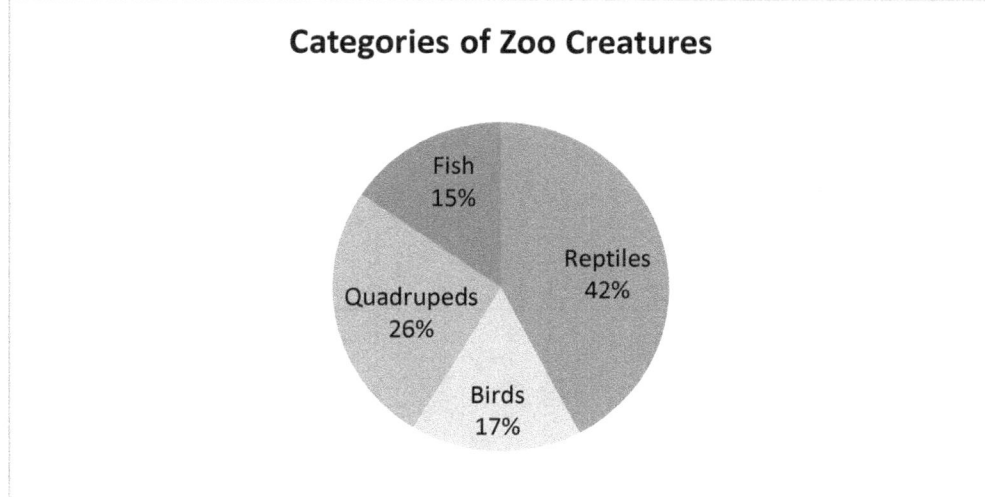

45
© COPYRIGHT 2017. Exam SAM Study Aids & Media dba www.examsam.com
This material may not be copied or reproduced in any form.

At the end of the year, the zoo still has 1,500 creatures, but reptiles constitute 40%, birds 23%, and quadrupeds 21%. How many more fish were there at the end of the year than at the beginning of the year?

A) 10
B) 11
C) 15
D) 16

> This question is asking you to interpret a pie chart that shows percentages by category. If you are asked to calculate changes to the data in the categories in the chart, be sure to multiply by the percentages at the beginning of the year and then do a separate calculation using the percentages at the end of the year.

35) A jar contains 4 red marbles, 6 green marbles, and 10 white marbles. If a marble is drawn from the jar at random, what is the probability that this marble is white?

A) 1/2
B) 1/5
C) 1/10
D) 3/10

36) Which of the following number lines represents seven values in which the median of the values exceeds the mean of the values?

A)

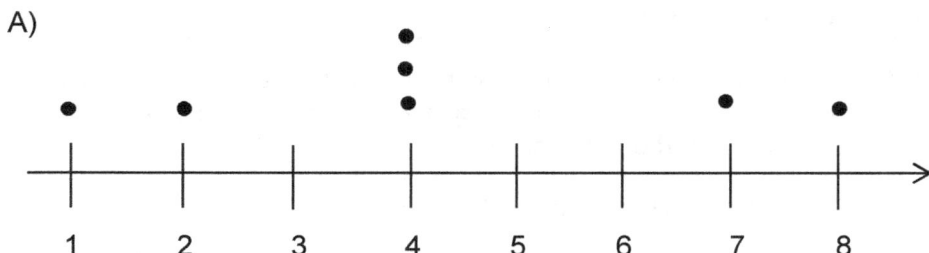

B)

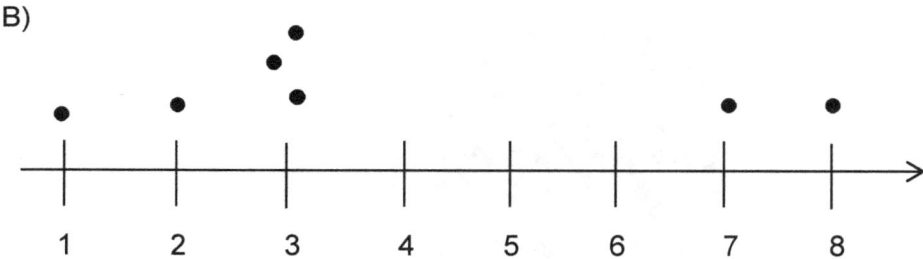

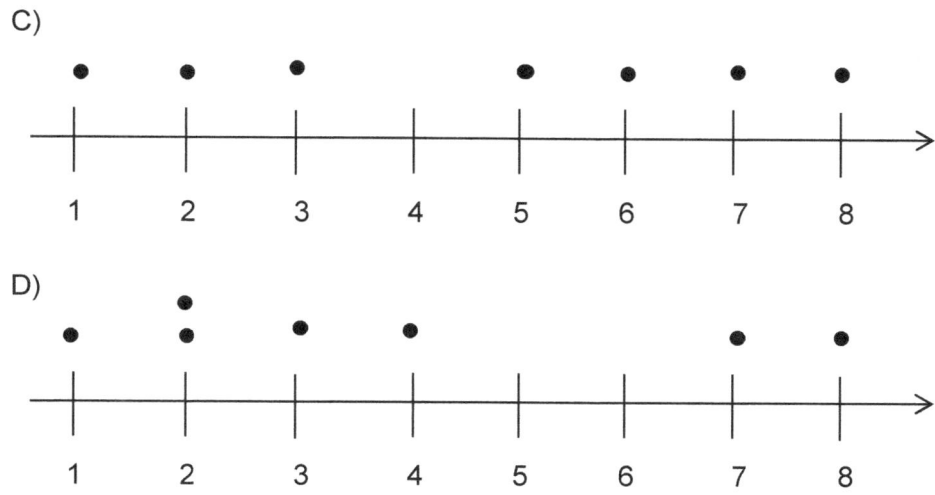

This is a question on reading number lines to interpret distributions. To determine the distribution, you should first look to see how many dots there are above the lines. Then add up the individual values from each line to calculate the mean of each of the five answer options. You can determine the median visually by seeing which number is midway on each of the lines.

Problem solving, reasoning, and mathematical communication:

37) Use the information in the box below to answer the question that follows.

> - School will be held every weekday from Monday to Friday from 15th August until 22nd December from 8:30 am to 3:00 pm.
> - However, if the temperature is more than 100 degrees, school will be dismissed at 11:30 am.
> - School will not be held on public holidays.

If Tom did not go to school today, then which of the following statements must be true?
A) It is the 22nd of August.
B) The temperature exceeds 100 degrees.
C) It is a public holiday or the temperature exceeds 100 degrees.
D) It is a public holiday or a Saturday or Sunday.

38) Read the problem below and answer the question that follows.

> - Dan rode his horse 2 miles to his neighbor's house.
> - It took the horse 15 minutes to make this journey.
> - From his neighbor's house, Dan rode his horse 3 miles into town.
> - What is the average pace of Dan's horse in miles per hour for these two journeys?

What piece of information is needed in order to solve the problem?
A) The distance from Dan's house into town.
B) The amount of time Dan stayed at his neighbor's house.
C) The length of the stride of Dan's horse.
D) The amount of time it took to go from the neighbor's house into town.

39) In order to make an estimate for the next week, a museum counts its visitors each day and rounds each daily figure up or down to the nearest 5 people. 104 people visit the museum on Monday, 86 people visit the museum on Tuesday, and 81 people visit the museum on Wednesday. Which figure below represents the best estimate of the amount of visitors to the museum for these three days for next week?
A) 260
B) 265
C) 270
D) 275

40) Use the information below to answer the question that follows.

- The baseball team practices every Tuesday and Friday.
- There will be no practice during the last full week of the month.
- There will be no practice in the event of rain.

If there is practice today, which of the following conclusions can be made?
A) It is the last full week of the month.
B) It is a Tuesday or it is not raining.
C) It is a Tuesday and it is raining.
D) It is a Tuesday or a Friday.

41) If the pattern below continues, which figure is next in this sequence?

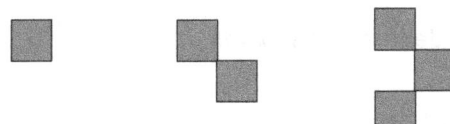

A)

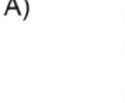

B)

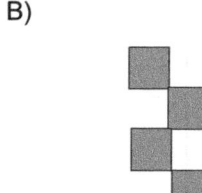

C)

D)

For questions on patterns that have illustrations this like one, look carefully at the layout of the figures. Here, we can see that the second block is placed at the lower right corner of the first block. So this pattern will also exist between the third block and the fourth block.

42) Use the diagram below to answer the question that follows.

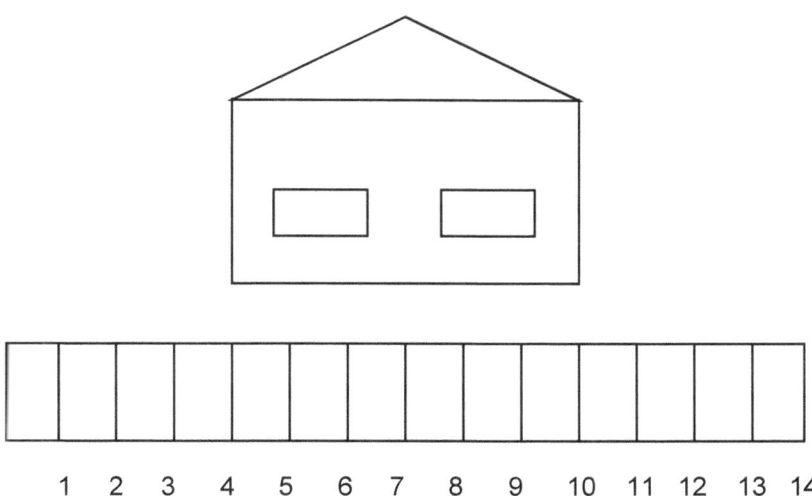

If each rectangle in the ruler below the picture of the house is one unit and the actual length of the house is 36 feet, then what is the scale of the diagram of the house?
A) 1 unit = 6 feet
B) 1 unit = 7.2 feet
C) 1 unit = 9 feet
D) 1 unit = 12 feet

For questions on scale drawings, just count the number of units below the figure on the ruler, rather than trying to subtract a certain amount of units from the total.

43) Jason does the high jump for his high school track and field team. His first jump is at 3.246 meters. His second is 3.331 meters, and his third is 3.328 meters. If the height of each jump is rounded to the nearest one-hundredth of a meter (also called a centimeter), what is the estimate of the total height for all three jumps combined?
A) 9.80
B) 9.89
C) 9.90
D) 9.91

44) Read the information below and answer the question that follows.

- If door A is locked with the red key, then door B is locked with the blue key.
- If door B is locked with the blue key, then door C is locked with the green key.
- The key that locks door B also locks door D.

If door A is locked with the red key, then which of the following must be true?
A) Door C is locked with the blue key.
B) Door C is locked with the red key.
C) Door D is locked with the blue key.
D) Door D is locked with the red key.

Read the facts in the problem carefully. It may also be helpful to draw a diagram or jot down some notes.

45) An amusement park gets 750,000 visitors a year, which is more than twenty thousand visitors per day.

Which of the following estimates correctly evaluates the reasonableness of the claim stated above?

A) reasonable, since 750,000 divided by 20,000 is approximately 365
B) reasonable, since 20,000 times 52 times 7 is more than 750,000
C) unreasonable, since 750,000 divided by 350 is closer to two thousand
D) unreasonable, 20,000 times 365 is less than 750,000

For questions on reasonableness of estimates, look at each answer choice and perform the operations indicated in each one.

Answer Key – NES Essential Academic Skills Practice Math Test 2

1) B	24) A
2) A	25) D
3) D	26) C
4) B	27) B
5) B	28) A
6) A	29) D
7) D	30) B
8) C	31) D
9) B	32) C
10) C	33) C
11) A	34) C
12) D	35) A
13) B	36) C
14) A	37) D
15) D	38) D
16) C	39) C
17) A	40) D
18) B	41) B
19) B	42) A
20) A	43) D
21) B	44) C
22) B	45) C
23) D	

Solutions and Explanations to NES Essential Academic Skills Practice Math Test 2

1) The correct answer is B. Remember to invert the second fraction by putting the denominator on the top and the numerator on the bottom. So the second fraction $\frac{4}{7}$ becomes $\frac{7}{4}$ when inverted. Multiply by the inverted fraction to solve the problem: $\frac{1}{5} \div \frac{4}{7} = \frac{1}{5} \times \frac{7}{4} = \frac{7}{20}$

2) The correct answer is A.

STEP 1: To find the LCD, you have to look at the factors for each denominator. Factors are the numbers that equal a product when they are multiplied by each other. So, the factors of 9 are:
1 × 9 = 9
3 × 3 = 9

The factors of 27 are:
1 × 27 = 27
3 × 9 = 27

STEP 2: Determine which factors are common to both denominators by comparing the lists of factors. In this problem, the factors of 3 and 9 are common to the denominators of both fractions. We can illustrate the common factors as shown below. We saw that the factors of 9 were:
1 × **9** = 9
3 × 3 = 9

The factors of 27 were:
1 × 27 = 27
3 × **9** = 27

So, the numbers in bold above are the common factors.

STEP 3: Multiply the common factors to get the lowest common denominator. The numbers that are in bold above are used to calculate the lowest common denominator: 3 × 9 = 27

STEP 4: Convert the denominator of each fraction to the LCD. You convert the fraction by referring to the factors from step 3. Multiply the numerator and the denominator by the same factor.

We convert the first fraction as follows: $\frac{1}{9} \times \frac{3}{3} = \frac{3}{27}$

We do not need to convert the second fraction of $\frac{9}{27}$ because it already has the LCD.

STEP 5: When both fractions have the same denominator, you can perform the operation to solve the problem: $\frac{1}{9} + \frac{9}{27} = \frac{3}{27} + \frac{9}{27} = \frac{12}{27}$

3) The correct answer is D.

STEP 1: Look at the factors of the numerator and denominator. The factors of 12 are:
1 × 12 = 12
2 × 6 = 12

3 × 4 = 12

You will remember that the factors of 27 are:
1 × 27 = 27
3 × 9 = 27

So, we can see that the numerator and denominator have the common factor of 3.

STEP 2: Simplify the fraction by dividing the numerator and denominator by the common factor.
Simplify the numerator: 12 ÷ 3 = 4
Then simplify the denominator: 27 ÷ 3 = 9

STEP 3: Use the results from step 2 to form the new fraction. The numerator from step 2 is 4. The denominator is 9. So, the new fraction is $\frac{4}{9}$

4) The correct answer is B. If $t\%$ subscribe to digital TV packages, then $100\% - t\%$ do not subscribe. In other words, since a percentage is any given number out of 100%, the percentage of students who do not subscribe is represented by this equation: $(100\% - t\%)$. This equation is then multiplied by the total number of students (n) in order to determine the number of students who do not subscribe to digital TV packages: $(100\% - t\%) \times n$

5) The correct answer is B. The sales price of each cell phone is four times the cost. The cost is expressed as x, so the sales price is $4x$. The difference between the sales price of each cell phone and the cost of each cell phone is the profit. In this problem, the sales price is $4x$ and the cost is x.

Sales Price − Cost = Profit
$4x − x$ = Profit
$3x$ = Profit

6) The correct answer is A. Remember the two concepts: (a) Negative numbers are less than positive numbers; (b) When two fractions have the same numerator, the fraction with the smaller number in the denominator is the larger fraction. So, $-1/4$ is less than $1/8$, $1/8$ is less than $1/6$, and $1/6$ is less than 1.

7) The correct answer is D. First of all, you need to determine the difference in temperature during the entire time period: 62 − 38 = 24 degrees less. Then calculate how much time has passed. From 5:00 PM to 11:00 PM, 6 hours have passed. Next, divide the temperature difference by the amount of time that has passed to get the temperature change per hour:
24 degrees ÷ 6 hours = 4 degrees less per hour

To calculate the temperature at the stated time, you need to calculate the time difference. From 5:00 PM to 9:00 PM, 4 hours have passed. So, the temperature difference during the stated time is 4 hours × 4 degrees per hour = 16 degrees less.

Finally, deduct this from the beginning temperature to get your final answer: 62° F − 16° F = 46° F

8) The correct answer is C. The number of hot dogs is D and the number of hamburgers is H. The equation to express the problem is: $(D \times \$2.50) + (H \times \$4) = \$22$
We know that the number of hamburgers is 3, so put that in the equation and solve it.
$(D \times \$2.50) + (H \times \$4) = \$22$
$(D \times \$2.50) + (3 \times \$4) = \$22$
$(D \times \$2.50) + 12 = \$22$

$(D \times \$2.50) + 12 - 12 = \$22 - 12$
$(D \times \$2.50) = \$10$
$\$2.50D = \$10$
$\$2.50D \div \$2.50 = \$10 \div \$2.50$
$D = 4$

9) The correct answer is B. The base number in this example is 10. So, we subtract the exponents: $10^6 \div 10^4 = 10^{(6-4)} = 10^2$

10) The correct answer is C. Isolate the integers to one side of the equation.

$\frac{3}{4}x - 2 = 4$

$\frac{3}{4}x - 2 + 2 = 4 + 2$

$\frac{3}{4}x = 6$

Then get rid of the fraction by multiplying both sides by the denominator.

$\frac{3}{4}x \times 4 = 6 \times 4$

$3x = 24$

Then divide to solve the problem.
$3x \div 3 = 24 \div 3$
$x = 8$

11) The correct answer is A.

Perform the multiplication on the terms in the parentheses.
$2(3x - 1) = 4(x + 1) - 3$
$6x - 2 = (4x + 4) - 3$

Then simplify.
$6x - 2 = (4x + 4) - 3$
$6x - 2 = 4x + 1$
$6x - 2 - 1 = 4x + 1 - 1$
$6x - 3 = 4x$

Then isolate x to get your answer.
$6x - 3 = 4x$
$6x - 4x - 3 = 4x - 4x$
$2x - 3 = 0$
$2x - 3 + 3 = 0 + 3$
$2x = 3$
$2x \div 2 = 3 \div 2$
$x = {}^3/_2$

12) The correct answer is D. Place the integers on one side of the inequality.
$-3x + 14 < 5$
$-3x + 14 - 14 < 5 - 14$

$-3x < -9$

Then get rid of the negative number. We need to reverse the way that the inequality sign points because we are dividing by a negative.

$-3x < -9$

$-3x \div -3 > -9 \div -3$ ("Less than" becomes "greater than" because we divide by a negative number.)

$x > 3$

4.35 is greater than 3, so it is the correct answer.

13) The correct answer is B. You need to find the lowest common denominator. Then add the numerators together as shown.

$$\frac{x}{5} + \frac{y}{2} =$$

$$\left(\frac{x}{5} \times \frac{2}{2}\right) + \left(\frac{y}{2} \times \frac{5}{5}\right) =$$

$$\frac{2x}{10} + \frac{5y}{10} =$$

$$\frac{2x + 5y}{10}$$

14) The correct answer is A. Multiply each side of the equation by Z. Then divide by W in order to isolate Z.

$$W = \frac{XY}{Z}$$

$$W \times Z = \frac{XY}{Z} \times Z$$

$$WZ = XY$$

$$WZ \div W = XY \div W$$

$$Z = \frac{XY}{W}$$

15) The correct answer is D. The price of the internet connection is always 5 times the speed.

10 = 2 × 5
20 = 4 × 5
30 = 6 × 5
40 = 8 × 5

So, the price of the internet connection (represented by variable P) equals the speed (represented by variable s) times 5: $P = s \times 5$

16) The correct answer is C. Study the solution below, which highlights the order to carry out the FOIL method to perform the operations on the terms.

$(3x - 2y)^2 = (3x - 2y)(3x - 2y)$

FIRST: The first terms in each set of parentheses are $3x$ and $3x$: $(\mathbf{3x} - 2y)(\mathbf{3x} - 2y)$
$3x \times 3x = 9x^2$

OUTSIDE: The terms on the outside are $3x$ and $-2y$: $(\mathbf{3x} - 2y)(3x - \mathbf{2y})$
$3x \times -2y = -6xy$

INSIDE: The terms on the inside are $-2y$ and $3x$: $(3x - \mathbf{2y})(\mathbf{3x} - 2y)$
$-2y \times 3x = -6xy$

LAST: The last terms in each set are $-2y$ and $-2y$: $(3x - \mathbf{2y})(3x - \mathbf{2y})$
$-2y \times -2y = 4y^2$

All of these individual results are put together for your final answer to the question.
$9x^2 - 6xy - 6xy + 4y^2 =$
$9x^2 - 12xy + 4y^2$

17) The correct answer is A. Any real number squared will always equal a positive number. Since 8 is added to the first value x^2, the result will always be 8 or greater. In other words, since x^2 is always a positive number, the result of the equation would never be 0. So, there are zero solutions for this equation.

18) The correct answer is B. First, find the midpoint of the x coordinates for $(\mathbf{-4}, 2)$ and $(\mathbf{2},-4)$.
midpoint $x = (x_1 + x_2) \div 2$
midpoint $x = (-4 + 2) \div 2$
midpoint $x = -2 \div 2$
midpoint $x = -1$

Then find the midpoint of the y coordinates for $(-4, \mathbf{2})$ and $(2,\mathbf{-4})$.
midpoint $y = (y_1 + y_2) \div 2$
midpoint $y = (2 + -4) \div 2$
midpoint $y = -2 \div 2$
midpoint $y = -1$
So, the midpoint is $(-1, -1)$

19) The correct answers is B. When a transversal crosses two parallel lines, opposite angles will be equal in measure and corresponding angles will also be equal in measure. (Corresponding angles are angles in the matching same-shaped corners.) Angles ∠b and ∠c are opposite angles and angles ∠c and ∠f are corresponding angles, so answer B is correct. Angles ∠b and ∠e are corresponding and angles ∠e and ∠f are also opposite.

20) The correct answer is A. We start off with point B, which is represented by the coordinates (0, 2). The line is then shifted 5 units to the left and 4 units up. When we go to the left, we need to deduct the units, and when we go up we need to add units. So, do the operations on each of the coordinates in order to solve: $0 - 5 = -5$ and $2 + 4 = 6$, so our new coordinates are $(-5, 6)$.

21) The correct answer is B. The perimeter is the measurement along the outside edges of the rectangle or other area. The formula for perimeter is as follows: $P = 2W + 2L$

If the room is 12 feet by 10 feet, we need 12 feet × 2 feet to finish the long sides of the room and 10 feet × 2 feet to finish the shorter sides of the room.

(2 × 10) + (2 × 12) =
20 + 24 = 44

22) The correct answer is B. Cone volume = (π × radius² × height) ÷ 3
Substitute the values for base and height.
volume = (π3² × 4) ÷ 3 =
(π9 × 4) ÷ 3 =
π36 ÷ 3 = 12π

23) The correct answer is D. In our problem we know that one side of the triangle is 18 meters and the other side of the triangle is 30 meters, so we can put these values into the formula in order to solve the problem.

$\sqrt{A^2 + B^2} = C$

$\sqrt{18^2 + 30^2} = C$

$\sqrt{324 + 900} = C$

$\sqrt{1224} = C$

35 × 35 = 1225

So, the square root of 1224 is approximately 35.

24) The correct answer is A. An isosceles triangle has two equal sides, so answer A is correct. If an altitude is drawn in an isosceles triangle, we have to put a straight line down the middle of the triangle from the peak to the base. Dividing the triangle in this way would form two right triangles, rather than two equilateral triangles. So, answer B is incorrect. The base of an isosceles triangle can be longer than the length of each of the other two sides, so answer C is incorrect. The sum of all three angles of any triangle must be 180 degrees, rather than 360 degrees. So, answer D is incorrect. By definition a triangle must have three sides. All three angles inside the triangle must add up to 180 degrees and right angles measure 90 degrees.

25) The correct answer is D. First, calculate the area of the central rectangle. Remember that the area of a rectangle is length times height: 8 × 3 = 24
Using the Pythagorean Theorem, we know that the base of each triangle is 4.

$5 = \sqrt{3^2 + base^2}$
$5^2 = 3^2 + base^2$
$25 = 9 + base^2$
$25 - 9 = 9 - 9 + base^2$
$16 = base^2$
$4 = base$

Then calculate the area of each of the triangles on each side of the central rectangle. Remember that the area of a triangle is base times height divided by 2: (4 × 3) ÷ 2 = 6

So, the total area is the area of the main rectangle plus the area of each of the two triangles.
24 + 6 + 6 = 36

26) The correct answer is C. Essentially a rectangle is missing at the upper left-hand corner of the figure. We would need to know both the length and width of the "missing" rectangle in order to calculate the area of our figure. So, we need to know both X and Y in order to solve the problem.

27) The correct answer is B. Triangle area is base times height divided by 2. First, calculate the area of triangle NKM: $[6 \times (8 + 10)] \div 2 = (6 \times 18) \div 2 = 108 \div 2 = 54$

Then, calculate the area of the area of triangle NKL: $(6 \times 8) \div 2 = 24$

The remaining triangle NLM is then calculated by subtracting the area of triangle NKL from triangle NKM: $54 - 24 = 30$

28) The correct answer is A. In this case, we have two items, each of which has a variable outcome. There are 6 numbers on the black die and 6 numbers on the red die. Using multiplication, we can see that there are 36 possible combinations: $6 \times 6 = 36$

To check your answer, you can list the possibilities of the various combinations:

(1,1)   (1,2)   (1,3)   (1,4)   (1,5)   (1,6)
(2,1)   (2,2)   (2,3)   (2,4)   (2,5)   (2,6)
(3,1)   (3,2)   (3,3)   (3,4)   (3,5)   (3,6)
(4,1)   (4,2)   (4,3)   (4,4)   (4,5)   (4,6)
(5,1)   (5,2)   (5,3)   **(5,4)**   (5,5)   (5,6)
(6,1)   (6,2)   (6,3)   (6,4)   (6,5)   (6,6)

If the number on the left in each set of parentheses represents the black die and the number on the right represents the red die, we can see that there is one chance that Becky will roll a 4 on the red die and a 5 on the black die.

The result is expressed as a fraction, with the event (chance of the desired outcome) in the numerator and the total sample space (total data set) in the denominator.

So, the answer is $1/36$.

29) The correct answer is D. You need to find the total points for all the females and the total points for all the males: Females: $60 \times 95 = 5700$; Males: $50 \times 90 = 4500$. Then add these two amounts together and divide by the total number of students in the class to get your solution: $(5700 + 4500) \div 110 = 10,200 \div 110 = 92.73$ average for all 110 students

30) The correct answer is B. The mean is the arithmetic average. First, add up all of the items: $-2\% + 5\% + 7.5\% + 14\% + 17\% + 1.3\% + -3\% = 39.8\%$. Then divide by 7 since there are 7 companies in the set: $39.8\% \div 7 = 5.68\% \approx 5.7\%$

31) The correct answer is D. The range is the highest number minus the lowest number. Our data set is: $65000, $52000, $125000, $89000, $36000, $84000, $31000, $135000, $74000, and $87000. So, the range is: $135000 - $31000 = $104000

32) The correct answer is C. For ten sandwiches, the total price is $85, so each sandwich sells for $8.50: $85 total sales in dollars ÷ 10 sandwiches sold = $8.50 each

33) The correct answer is C. The quantity of diseases is indicated on the bottom of the graph, while the number of children is indicated on the left side of the graph. To determine the amount of children that have been vaccinated against three or more diseases, we need to add the amounts represented by the bars for 3, 4, and 5 diseases: 30 + 20 + 10 = 60 children

34) The correct answer is C. At the beginning of the year, 15% of the 1,500 creatures were fish, so there were 225 fish at the beginning of the year (1,500 × 0.15 = 225).

In order to find the percentage of fish at the end of the year, we first need to add up the percentages for the other animals: 40% + 23% + 21% = 84%

Then subtract this amount from 100% to get the remaining percentage for the fish: 100% − 84% = 16%

Multiply the percentage by the total to get the number of fish at the end of the year: 1,500 × 0.16 = 240

Then subtract the beginning of the year from the end of the year to calculate the increase in the number of fish: 240 − 225 = 15

35) The correct answer is A. Your first step is to calculate the total amount of items in the data set: 4 red marbles + 6 green marbles + 10 white marbles = 20 marbles in total. The probability is expressed with the subset in the numerator and the total remaining data set in the denominator. So, the chance of drawing a white marble is $^{10}/_{20} = ^{1}/_{2}$

36) The correct answer is C. The mean is the average of all of the numbers in the set. If we look at each of the answers, we can see that we have seven values in each set because there are seven dots above each of the number lines.

The mean for answer choice C is 4.57 and the median is 5.

Mean: 1 + 2 + 3 + 5 + 6 + 7 + 8 = 32; 32 ÷ 7 = 4.57

Median: 1, 2, 3, **5**, 6, 7, 8

So, the median exceeds the mean for the set represented on number line (C).

37) The correct answer is D. This is another type of problem in which you have to assess the available facts, so read carefully and do not make any assumptions that are not supported by the information provided. As stated previously, it is usually best to deal with each answer option one by one for these types of questions.

Answer choice A is incorrect. If it is the 22nd of August, Tom would be in school because school is held every weekday from Monday to Friday from 15th August until 22nd December.

Answer choices B and C are incorrect because even if the temperature was in excess of 100 degrees, Tom would have attended school from 8:30 to 11:30 am.

Therefore, we know that answer D is the correct answer. We also know that D is correct because school is held only on weekdays, and answer D stipulates that it is a Saturday, Sunday, or public holiday.

38) The correct answer is D. To calculate the average pace or speed, we need to know the speed for each journey. In order to calculate the speed of travel, we need to know the distance traveled and the amount of time for the journey. The problem tells us that Dan rode his horse 2 miles to his neighbor's house and that it took 15 minutes for this journey. So, we have both the distance traveled and the amount of time for the first journey. The problem also states that Dan made a second journey, riding his horse 3 miles into town from his neighbor's house. So, we have the distance traveled for the second journey, but we do not have the amount of time for the second journey.

39) The correct answer is C. If you look at the answer choices, you will see that they are given in the nearest increments of 5. So, we have to round the figures stated in the problem up or down to the nearest multiple of 5.

104 on Monday is rounded to 105.

86 on Tuesday is rounded down to 85.

81 is rounded down to 80.

Then add these three figures together to get your result.

105 + 85 + 80 = 270

40) The correct answer is D. If there is practice today, we can conclude that it is a Tuesday or a Friday. The facts tell us that there will be no practice during the last full week of the month and that there will be no practice in the event of rain.

41) The correct answer is B. The second block is placed at the lower right corner of the first block.

This pattern will also exist between the third block and the fourth block. So, repeat the pattern.

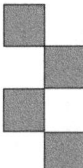

42) The correct answer is A. Count the number of units that the house spans, rather than trying to subtract units from the total of 14. If we count the number of units below the house in the drawing, we can see that the house spans 6 units. Divide this result into the actual length of the house (36 feet) to get the scale of the drawing. 36 feet ÷ 6 units = 6 feet represented by each unit

43) The correct answer is D. We know that we have to round to the nearest hundredth. The hundredth decimal place is the number 2 positions to the right of the decimal. For example, .01 is 1 one hundredth.

In our question, the first jump of 3.246 is rounded up to 3.25

The second jump of 3.331 is rounded down to 3.33

The third jump of 3.328 is rounded up to 3.33

Then add these three figures together to get your answer: 3.25 + 3.33 + 3.33 = 9.91

44) The correct answer is C. If door A is locked with the red key, then door B is locked with the blue key. If the key that locks door B also locks door D, then the blue key is also used to lock door D.

45) The correct answer is C. Perform the operations to determine which answer choice is correct. 750,000 ÷ 350 = 2,143, so answer C is correct.

NES Essential Academic Skills Practice Math Test 3

Number properties and number operations:

1) The numbers in the following list are ordered from greatest to least: Θ, η, $^{25}/_{13}$, $^{10}/_{9}$, $^{1}/_{3}$. Which of the following could be the value of η?
   A) $\sqrt{36}$
   B) $^{25}/_{14}$
   C) $^{24}/_{13}$
   D) 1.91

2) If $7x$ is between 5 and 6, which of the following could be the value of x?
   A) $^{2}/_{3}$
   B) $^{3}/_{4}$
   C) $^{5}/_{8}$
   D) $^{7}/_{8}$

3) A painter needs to paint 8 rooms, each of which have a surface area of 2000 square feet. If one bucket of paint covers 900 square feet, what is the fewest number of buckets of paint that must be used to complete all 8 rooms?
   A) 3
   B) 17
   C) 18
   D) 19

4) The price of a certain book is reduced from $60 to $45 at the end of the semester. By what percent is the price of the book reduced?
   A) 15%
   B) 20%
   C) 25%
   D) 33%

5) The ratio of males to females in the senior year class of Carson Heights High School was 6 to 7. If the total number of students in the class is 117, how many males are in the class?
   A) 48
   B) 54
   C) 56
   D) 58

6) Shanika works as a car salesperson. She earns $1,000 a month, plus $390 for each car she sells. If she wants to earn at least $4,000 this month, what is the minimum number of cars that she must sell this month?
   A) 6
   B) 7
   C) 8
   D) 9

7) Which one of the values will correctly satisfy the following mathematical statement:

$$\frac{2}{3} < ? < \frac{7}{9}$$

A) $\frac{1}{3}$
B) $\frac{1}{5}$
C) $\frac{2}{6}$
D) $\frac{7}{10}$

8) The ratio of bags of apples to bags of oranges in a particular grocery store is 2 to 3. If there are 44 bags of apples in the store, how many bags of oranges are there?
A) 33
B) 48
C) 55
D) 66

9) At the beginning of class, $\frac{1}{5}$ of the students leave to go to singing lessons. Then $\frac{1}{4}$ of the remaining students leave to go to the principal's office. If 18 students are then left in the class, how many students were there at the beginning of class?
A) 90
B) 45
C) 30
D) 25

Algebra and graphing:

10) The graph of $y = 8 \div (x - 4)$ is shown below.

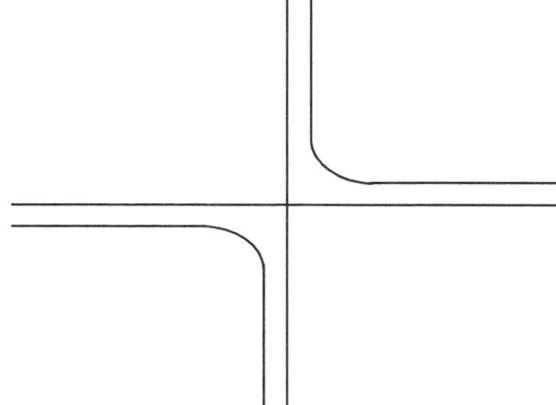

Which of the following is the best representation of $8 \div |(x - 4)|$?

A)

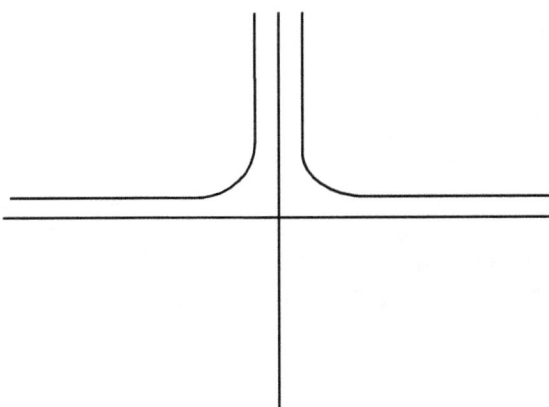

B)

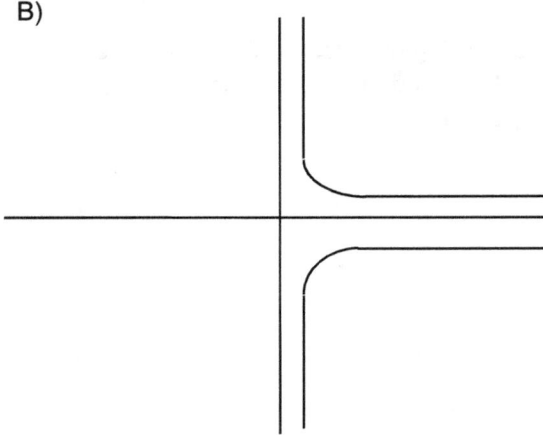

C)

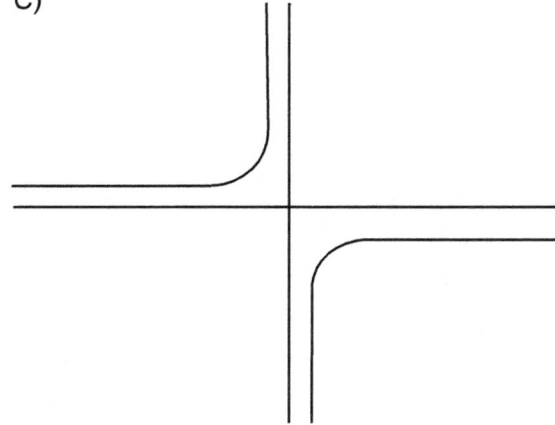

D)

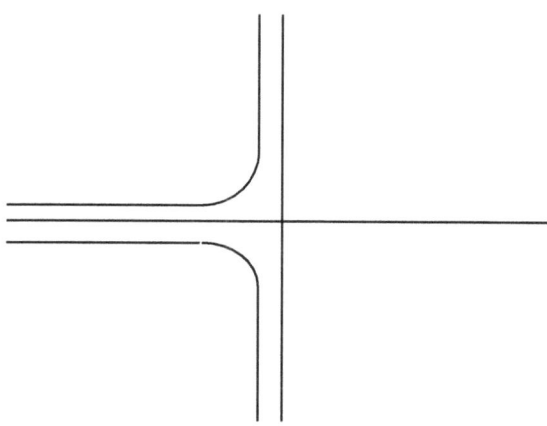

11) Which of the following shows the solution set of −3x > 6?

A)
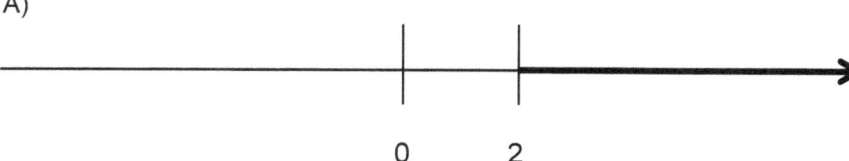

B)
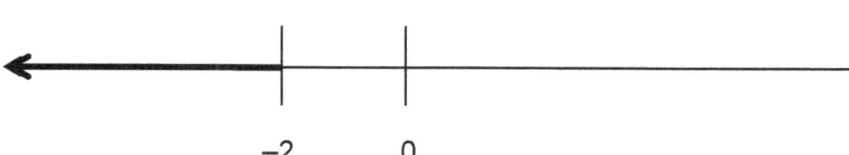

C)
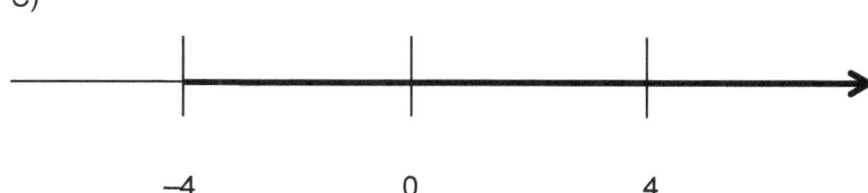

D)
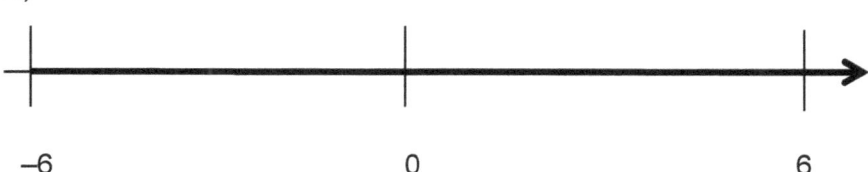

12) Fatima drove into town at a rate of 50 miles per hour. She shopped in town for 20 minutes, and then drove home on the same route at a rate of 60 miles per hour. Which of the following equations best expresses the total time (Tt) that it took Fatima to make the journey and do the shopping? Note that the variable D represents the distance in miles from Fatima's house to town.
A) $Tt + 20 \text{ minutes} = 110 \times D$
B) $Tt + 20 \text{ minutes} = [(50 + 60) \div 2] \times D$
C) $Tt = [(D \div 50) + (D \div 60)] + 20 \text{ minutes}$
D) $Tt = D \div 110$

13) A baseball team sells T-shirts and sweatpants to the public for a fundraising event. The total amount of money the team earned from these sales was $850. Variable t represents the number of T-shirts sold and variable s represents the number of sweatpants sold. The total sales in dollars is represented by the equation $25t + 30s$. What equation represents the fraction of the amount earned by selling sweatpants to the total amount earned?
A) s/850
B) 30s/850
C) (25t + 30s)/850
D) t/850

14) A company is making its budget for the cost of employees to attend conferences for the year. It costs $7,500 per year in total for the company plus C dollars per employee. During the year, the company has E employees. If the company has budgeted $65,000 for conference attendance, which equation can be used to calculate the maximum cost per employee?
A) $(\$65{,}000 - \$7{,}500) \div E$
B) $(\$65{,}000 - \$7{,}500) \div C$
C) $(C - \$7{,}500) \div E$
D) $\$65{,}000 \div E$

15) A table of values for a linear equation is given below. What is the missing value of y?

x	y
2	4
4	6
6	8
8	?
10	12

A) 7
B) 8
C) 9
D) 10

16) Find the coordinates (x, y) of the midpoint of the line segment on a graph that connects the points (–5, 2) and (7, 2).
A) (1, 2)
B) (2, 1)
C) (1, –2)
D) (2, 2)

17) The graph of a line is shown on the xy plane below. The point that has the y-coordinate of 45 is not shown. What is the corresponding x-coordinate of that point?

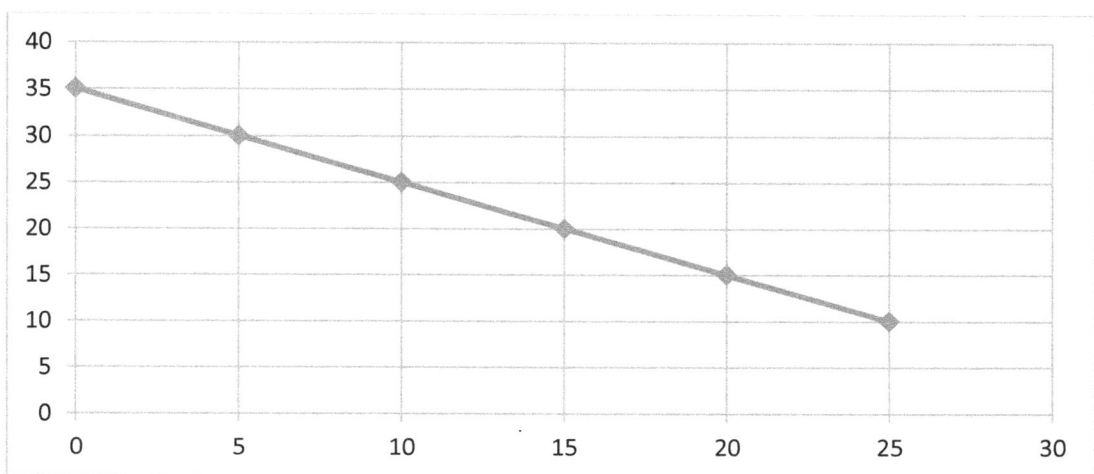

A) –10
B) –5
C) 0
D) 5

18) The graph of a line is shown on the xy plane below. The point that has the x-coordinate of 160 is not shown. What is the corresponding y-coordinate of that point?

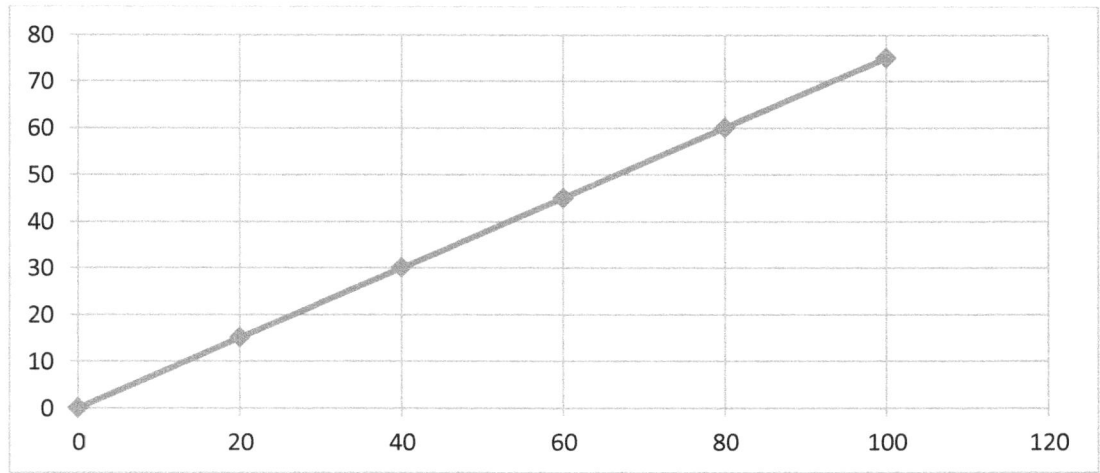

A) 115
B) 120
C) 125
D) 130

Geometry and measurement:

19) The area of a square is 64 square units. This square is made up of smaller squares that measure 4 square units each. How many of the smaller squares are needed to make up the larger square?
    A) 8
    B) 12
    C) 16
    D) 24

20) The perimeter of the square shown below is 24 units. What is the length of line segment AB?

    A
    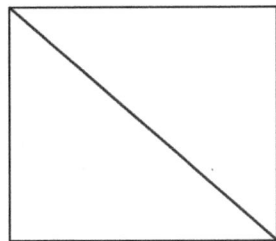
    B

    A) $\sqrt{24}$
    B) $\sqrt{36}$
    C) $\sqrt{72}$
    D) 6

21) Soon Li jogged 3.6 miles in $3/4$ of an hour. What was her average jogging speed in miles per hour?
    A) 2.7
    B) 4.0
    C) 4.2
    D) 4.8

22) ∠XYZ is an isosceles triangle, where XY is equal to YZ. Angle Y is 30° and points W, X, and Z are co-linear. What is the measurement of ∠WXY?

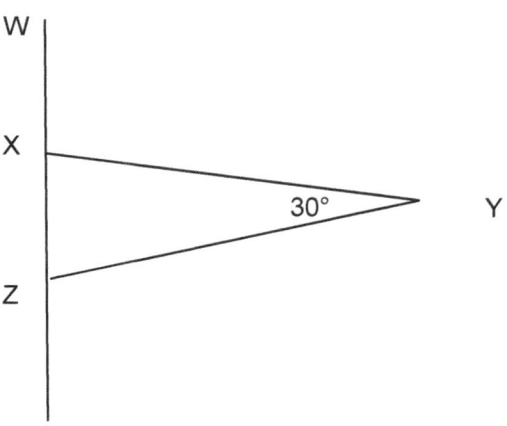

A) 40
B) 105
C) 150
D) 160

23) The Johnson's have decided to remodel their upstairs. They currently have 4 rooms upstairs that measure 10 feet by 10 feet each. When they remodel, they will make one large room that will be 20 feet by 10 feet and two small rooms that will each be 10 feet by 8 feet. The remaining space is to be allocated to a new bathroom. What are the dimensions of the new bathroom?
A) 4 × 10
B) 8 × 10
C) 10 × 10
D) 4 × 8

24) The line on the xy-graph below forms the diameter of the circle. What is the approximate circumference of the circle?

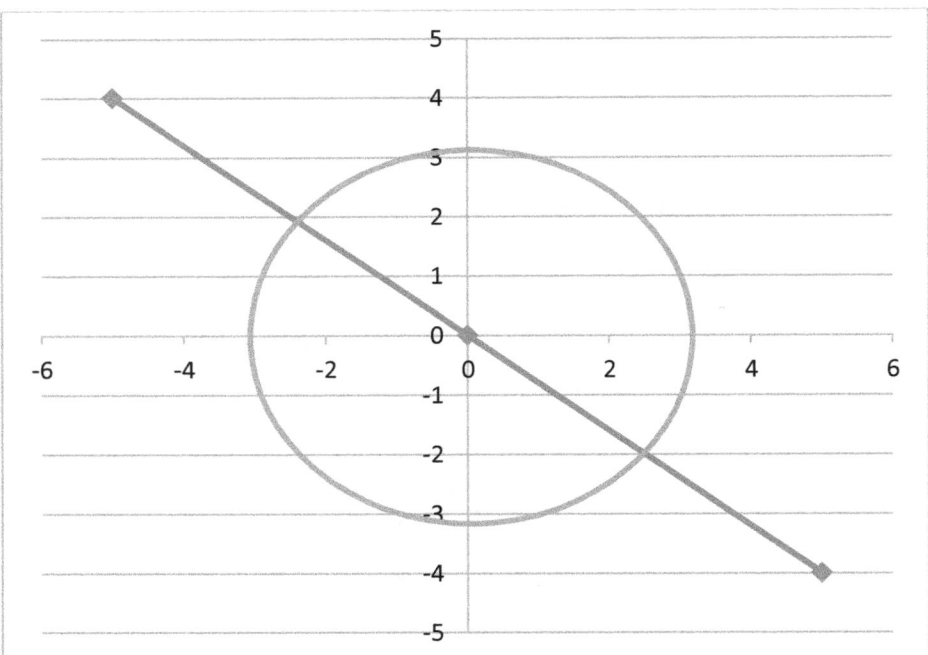

A) 3π
B) 6
C) 6π
D) 9

25) If circle A has a radius of 0.4 and circle B has a radius of 0.2, what is the difference in area between the two circles?

A) $.04\pi$
B) $.12\pi$
C) $.16\pi$
D) $.40\pi$

26) Which of the following statements about parallelograms is false?
A) A parallelogram has no right angles.
B) A parallelogram has opposite angles which are congruent.
C) A parallelogram has only one pair of parallel sides.
D) The opposite sides of a parallelogram are unequal in measure.

27) Which of the following statements best describes supplementary angles?
A) Supplementary angles must add up to 90 degrees.
B) Supplementary angles must add up to 180 degrees.
C) Supplementary angles must add up to 360 degrees.
D) Supplementary angles must be congruent angles.

Probability and statistics:

28) A student receives the following scores on his exams during the semester:
   89, 65, 75, 68, 82, 74, 86
   What is the mean of his scores?
   A) 24
   B) 74
   C) 75
   D) 77

29) A clown pulls balloons out of a bag at random to blow up and give to children during a birthday party. At the start of the party, there are 10 red balloons, 7 green balloons, 6 purple balloons, 5 orange balloons, and 11 blue balloons in the bag. The clown selects the first balloon, which is blue, and gives it to the first child. If the second child gets an orange balloon, what is the probability that the third child will get a blue balloon?
   A) $11/37$
   B) $10/37$
   C) $11/39$
   D) $10/39$

30) What is the mode of the numbers in the following list?
   1.6, 2.9, 4.5, 2.5, 2.5, 5.1, 5.4
   A) 3.5
   B) 3.1
   C) 3.0
   D) 2.5

31) There are 10 cars in a parking lot. Nine of the cars are 2, 3, 4, 5, 6, 7, 9, 10, and 12 years old, respectively. If the average age of the 10 cars is 6 years old, how old is the 10th car?
   A) 1 year old
   B) 2 years old
   C) 3 years old
   D) 4 years old

32) Which of the following is a statistical question?
   A) What size shoe does Mrs. Shapiro wear?
   B) How many residents of the town oppose the tax increase?
   C) Will it rain tomorrow?
   D) How many miles can that car travel on a tank of gasoline?

33) The range and mean of 5 numbers are 10 and 14 respectively. The five numbers are positive integers greater than 0. If the range is increased by 2, which of the following could be true of the numbers in the new set? You may select more than one answer.
   A) The highest number is increased by 2.
   B) The highest number is decreased by 1 and the lowest number is decreased by 1.

C) The highest number is decreased by 1 and the lowest number is increased by 1.
D) The lowest number is increased by 2.

34) The students at Lyndon High School have been asked about their plans to attend the Homecoming Dance. The chart below shows the responses of each grade level by percentages. Which figure below best approximates the percentage of the total number of students from all four grades who will attend the dance? Note that each grade level has roughly the same number of students.

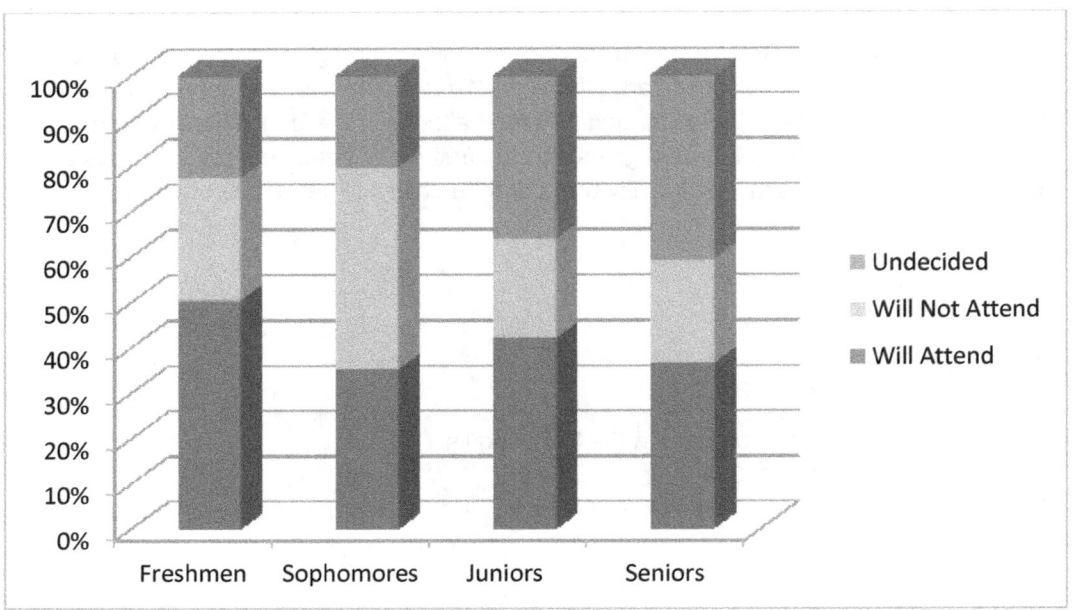

A) 25%
B) 35%
C) 45%
D) 55%

35) Data on the number of vehicles involved in traffic accidents in Cedar Valley on certain dates is represented in the chart below.

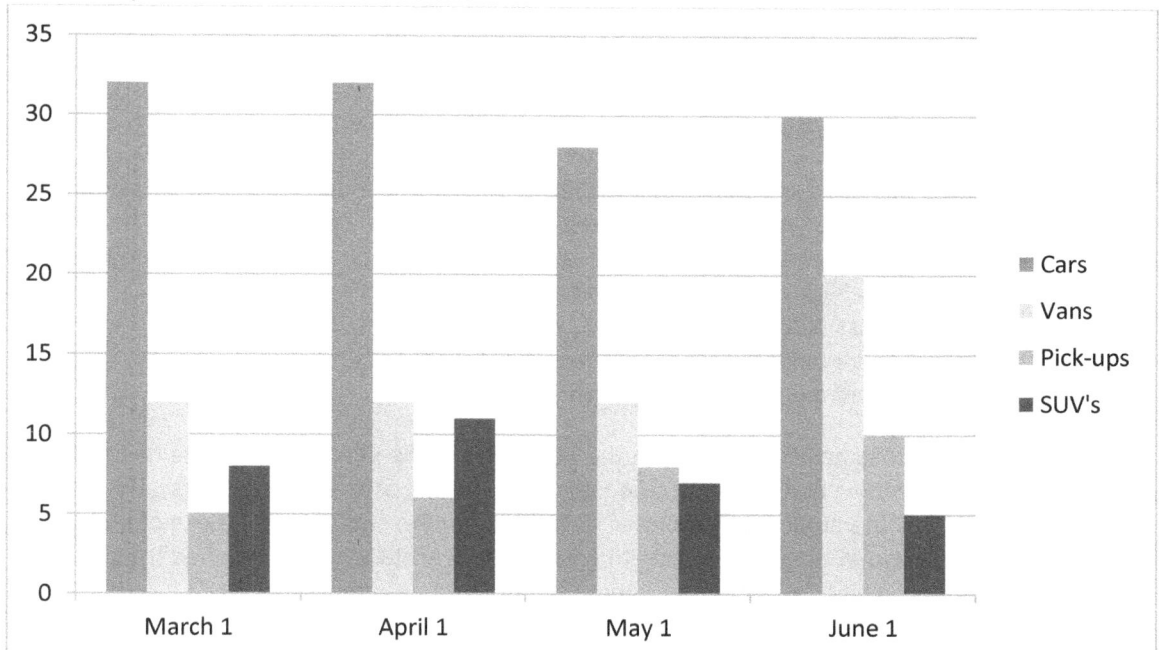

Pick-ups and vans were involved in approximately what percentage of total vehicle accidents on June 1?
A) 7.6%
B) 15%
C) 31%
D) 46%

36) The pictograph below illustrates the results of a customer satisfaction survey by region. Each of the four regions has one salesperson. Salespeople in each region receive bonuses based on the amount of positive customer feedback they receive. If the salespeople from all four regions received $540 in bonuses in total, how much bonus money does the company pay each individual salesperson per satisfied customer?

Region 1	☺ ☺ ☺ ☺
Region 2	☺ ☺ ☺
Region 3	☺ ☺
Region 4	☺ ☺ ☺

Each ☺ represents positive feedback from 10 customers.

A) $4.00
B) $4.50
C) $4.90
D) $5.00

Problem solving, reasoning, and mathematical communication:

37) There are three boys in a family, named Alex, Burt, and Zander. Alex is twice as old as Burt, and Burt is one year older than three times the age of Zander. Which of the following statements best describes the relationship between the ages of the boys?
A) Alex is 4 years older than 6 times the age of Zander.
B) Alex is 2 years older than 6 times the age of Zander.
C) Alex is 4 years older than 3 times the age of Zander.
D) Alex is 2 years older than 3 times the age of Zander.

38) A martial arts class has 53 students at the beginning of the year. 15 students have black belts, 22 have brown belts, 8 have blue belts, and 8 have belts of other colors. By the end of the year, 3 of the students with brown belts and 2 of the students with belts of other colors have dropped out of the class. In addition, 4 new students have joined the class. Which of the following facts can be determined from the information above?
A) The total number of students in the class.
B) The number of students in the class with brown belts.
C) The number of students in the class with blue belts.
D) The number of students in the class with black belts.

39) Use the table below to answer the question that follows.

Regional Railway Train Service	
Departure Time	Arrival Time
9:50 am	10:36 am
11:15 am	12:01 pm
12:30 pm	1:16 pm
2:15 pm	3:01 pm
?	5:51 pm

The journey on the Regional Railway is always exactly the same duration.
What is the missing time in the chart above?
A) 3:30 pm
B) 4:15 pm
C) 4:30 pm
D) 5:05 pm

40) Read the information in the box below and answer the question that follows:
- A health and beauty store has 90 bottles of shampoo for sale when the store opens for business on Monday morning.

- These 90 bottles of shampoo consist of 15 bottles of strawberry-scented shampoo, 25 bottles of rose-scented shampoo, and 50 bottles of unscented shampoo.
- At the close of business on Monday, 18 bottles of rose-scented shampoo remain in the store.

Which of the following facts can be determined from the information above?
A) The quantity of shampoo that the store normally offers for sale.
B) The average price of a bottle of shampoo.
C) The quantity of strawberry-scented shampoo sold on Monday.
D) The quantity of rose-scented shampoo sold on Monday.

41) Read the problem below and then answer the question that follows.
- Tom and Mary are planning a cross-country trip.
- They plan to drive 300 miles each day for seven days.
- Their car can travel 25 miles on one gallon of gasoline.
- How much money in total will they need to pay for gasoline during their trip?

What piece of information is needed in order to answer the problem?
A) The amount of gasoline that the tank of the car can hold.
B) The total amount of miles that they will drive that week.
C) The price per gallon of gasoline.
D) The day of the week that their journey will begin.

42) Read the problem below and then answer the question that follows.
- Paul leaves his house at 5:30 to go running.
- He runs 2 miles north through town, then continues 3 miles north out of town.
- He then runs south to his house along the same route.
- What is Paul's running pace?

What piece of information is needed in order to answer the problem?
A) The amount of steps that Paul makes.
B) The time that Paul returns home.
C) The length of Paul's stride.
D) The length of the return journey.

43) Use the information below to answer the question that follows.
- If the distance from his house to his destination is 5 miles or more, Jose uses his motorcycle.
- If the distance from his house to his destination is less than 5 miles but more than 1 mile, Jose uses his bicycle.
- If the distance from his house to his destination is 1 mile or less, Jose walks.

Jose uses his bicycle to go to Manuel's house. Which one of the following statements could be true?
A) Manuel's house is 1 mile or less from Jose's house.
B) Manuel and Jose live 8 miles apart.
C) Jose's house is at least 6 miles from Manuel's.
D) Jose lives 4 miles from Manuel.

44) A plumber charges $100 per job, plus $25 per hour worked. He is going to do 5 jobs this month. He will earn a total of $4,000. How can we calculate the amount of hours will he work this month?
   A) Subtract 100 from 4,000. Then divide this amount by 5.
   B) Subtract 100 from 4,000. Then divide this amount by 25.
   C) Multiply 100 by 5. Then subtract this amount from 4,000 to get a second amount. Then divide this second amount by 25.
   D) Divide 4,000 by 5. Then subtract 100 from this amount to get a second amount. Then divide this second amount by 25.

45) Use the following instructions to draw a diagram:

   I. Draw a line to the right 8 units long.
   II. Turn to the right 45 degrees.
   III. Then draw a line 8 units long.
   IV. Turn to the left 45 degrees.
   V. Then draw a line 8 units long.

   A)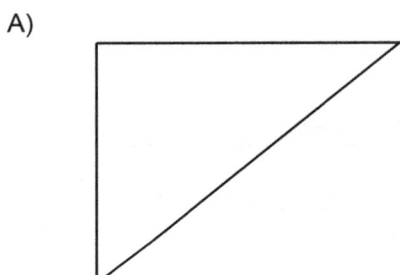

   B)

C)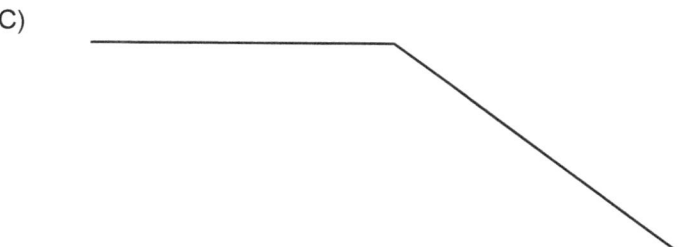

D)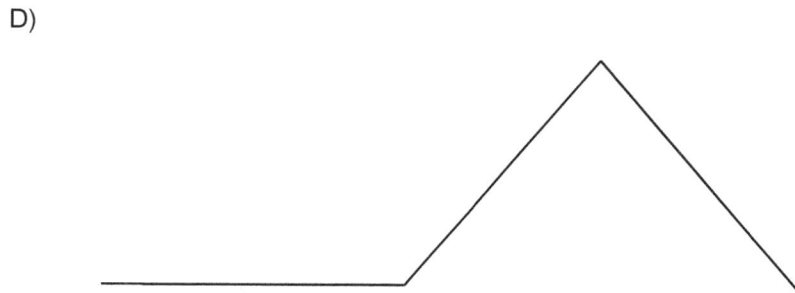

Answer Key – NES Essential Academic Skills Practice Math Test 3

1) A
2) B
3) C
4) C
5) B
6) C
7) D
8) D
9) C
10) A
11) B
12) C
13) B
14) A
15) D
16) A
17) A
18) B
19) C
20) C
21) D
22) B
23) A
24) C
25) B
26) B
27) B
28) D
29) B
30) D
31) B
32) B
33) A
34) B
35) D
36) B
37) B
38) A
39) D
40) D
41) C
42) B
43) D
44) C
45) C

Answers and Solutions to NES Essential Academic Skills Practice Math Test 3

1) The correct answer is A. From the facts in the problem, we know that η needs to be greater than $^{25}/_{13}$. If we convert $^{25}/_{13}$ to decimal form, we get 1.923077. The square root of 36 is 6, so (A) is the correct response because it is greater than 1.923077.

2) The correct answer is B. 7x is between 5 and 6, so set up an inequality as follows:
$5 < 7x < 6$

Then insert the fractions from the answer choices for the value of x to solve the problem.
$5 < (7 \times ^3/_4) < 6$
$5 < [(7 \times 3) \div 4] < 6$
$5 < (21 \div 4) < 6$
$5 < 5.25 < 6$
5.25 is between 5 and 6, so $^3/_4$ is the correct answer.

3) The correct answer is C. For your first step, determine how many square feet there are in total: 2000 square feet per room × 8 rooms = 16,000 square feet in total

Then you need to divide by the coverage rate:
16,000 square feet to cover ÷ 900 square feet coverage per bucket = 17.77 buckets needed
It is not possible to purchase a partial bucket of paint, so 17.77 is rounded up to 18 buckets of paint.

4) The correct answer is C. Determine the dollar amount of the reduction or discount:
$60 original price – $45 sale price = $15 discount

Then divide the discount by the original price to get the percentage of the discount:
$15 ÷ $60 = 0.25 = 25%

5) The correct answer is B. For your first step, add the subsets of the ratio together: 6 + 7 = 13
Then divide this into the total: 117 ÷ 13 = 9. The multiply by the ratio of males to get the amount of male students: 6 × 9 = 54

6) The correct answer is C. Shanika wants to earn $4,000 this month. She gets the $1,000 basic pay regardless of the number of cars she sells, so we need to subtract that from the total first:
$4,000 – $1,000 = $3,000

She gets $390 for each car she sells, so we need to divide that into the remaining $3,000:
$3,000 to earn ÷ $390 per car = 7.69 cars to sell

Since it is not possible to sell a part of a car, we need to round up to 8 cars.

7) The correct answer is D. First of all, we need to find a common denominator for the fractions in the inequality, as well as for the fractions in the answer choices. In order to complete the problem quickly, you should not try to find the lowest common denominator, but just find any common denominator. We can do this by expressing all of the numbers with a denominator of 90, since 9 is the largest denominator in the equation and 10 is the largest denominator in the answer choices.

$^2/_3 \times {^{30}/_{30}} = {^{60}/_{90}}$
$^7/_9 \times {^{10}/_{10}} = {^{70}/_{90}}$

Then, express the original equation in terms of the common denominator: $^{60}/_{90} < ? < ^{70}/_{90}$

Next, convert the answer choices to the common denominator.

A. $^{1}/_{3} \times ^{30}/_{30} = ^{30}/_{90}$
B. $^{1}/_{5} \times ^{18}/_{18} = ^{18}/_{90}$
C. $^{2}/_{6} \times ^{15}/_{15} = ^{30}/_{90}$
D. $^{7}/_{10} \times ^{9}/_{9} = ^{63}/_{90}$

Finally, compare the results to find the answer. By comparing the numerators (the top numbers of the fractions), we can see that $^{63}/_{90}$ lies between $^{60}/_{90}$ and $^{70}/_{90}$. So, D is the correct answer because $^{60}/_{90} < ^{63}/_{90} < ^{70}/_{90}$.

8) The correct answer is D. The ratio of bags of apples to bags of oranges is 2 to 3, so for every two bags of apples, there are three bags of oranges. First, take the total amount of bags of apples and divide by 2: 44 ÷ 2 = 22. Then multiply this by 3 to determine how many bags of oranges are in the store: 22 × 3 = 66.

9) The correct answer is C. Work backwards based on the facts given. There are 18 students left at the end after one-fourth of them left for the principal's office. So, set up an equation for this:

$18 + ^{1}/_{4}T = T$
$18 + ^{1}/_{4}T - ^{1}/_{4}T = T - ^{1}/_{4}T$
$18 = ^{3}/_{4}T$
$18 \times 4 = ^{3}/_{4}T \times 4$
$72 = 3T$
$72 \div 3 = 3T \div 3$
$24 = T$

So, before the group of pupils left to see the principal, there were 24 students in the class. We know that one-fifth of the students left at the beginning to go to singing lessons, so we need to set up an equation for this:

$24 + ^{1}/_{5}T = T$
$24 + ^{1}/_{5}T - ^{1}/_{5}T = T - ^{1}/_{5}T$
$24 = ^{4}/_{5}T$
$24 \times 5 = ^{4}/_{5}T \times 5$
$120 = 4T$
$120 \div 4 = 4T \div 4$
$30 = T$

10) The correct answer is A. We know from the original graph in the question that when x is a positive number, then y will also be positive. This is represented by the curve in the upper right-hand quadrant of the graph.

We also know from the original graph in the question that when x is negative, y will also be negative. This is represented by the curve in the lower left-hand quadrant of the graph.

If we add the absolute value symbols to the problem, then $|(x - 4)|$ will always result in a positive value for y.

Therefore, even when x is negative, y will be positive.

So, the curve originally represented in the lower left-hand quadrant of the graph must be shifted into the upper left-hand quadrant.

11) The correct answer is B.

Isolate the unknown variable in order to solve the problem.

$-3x > 6$

$-3x \div 3 > 6 \div 3$

$-x > 2$

In order to solve the problem, we have to multiply each side of the equation by -1.

When we multiply both sides of an inequality by a negative number, we have to reverse the greater than symbol to a less than symbol (or if there is a less than symbol, we reverse it to a greater than symbol).

$-x \times -1 < 2 \times -1$

$x < -2$

This is represented by line B.

12) The correct answer is C. The amount of time in hours (T) multiplied by miles per hour (mph) gives us the distance traveled (D). So, the equation for distance traveled is: $T \times mph = D$

The problem tells us that we need to calculate T, so we need to isolate T by changing our equation as follows:
$T \times mph = D$
$(T \times mph) \div mph = D \div mph$
$T = D \div mph$

In our problem, Fatima drives home on the same route that she took into town, so we need to calculate the traveling time for the journey into town, as well as for the journey home:
$(D \div 50) + (D \div 60) = T$

Then add back the 20 minutes she spent in town to get the total time:
$Tt = [(D \div 50) + (D \div 60)] + 20 \text{ minutes}$

13) The correct answer is B. We need to set up a fraction, the numerator of which consists of the amount of sales in dollars for sweatpants, and the denominator of which consists of the total amount of sales in dollars for both items. The problem tells us that the amount of sales in dollars for sweatpants is $30s$ and the total amount of sales is 850, so the answer is $30s/850$.

14) The correct answer is A. The total amount of the budget is $65,000. The up-front cost is $7,500, so we can determine the remaining amount of available funds by deducting the up-front cost from the total: $65,000 − $7,500. We have to divide the available amount by the number of employees (E) to get the maximum cost per employee: $(\$65,000 − \$7,500) \div E$

15) The correct answer is D. We need to look for the relationship between x and y. We can see that x + 2 is equal to y. So, add 2 to x to get the missing value for y: 8 + 2 = 10

16) The correct answer is A. Remember that in order to find midpoints on a line, you need to use the midpoint formula. For two points on a graph (x_1, y_1) and (x_2, y_2), the midpoint is:

$(x_1 + x_2) \div 2 , (y_1 + y_2) \div 2$
$(-5 + 7) \div 2$ = midpoint x, $(2 + 2) \div 2$ = midpoint y
$2 \div 2$ = midpoint x, $4 \div 2$ = midpoint y
1 = midpoint x, 2 = midpoint y

17) The correct answer is A. As x decreases by 5, y increases by 5. So, if we want to determine the x coordinate for $(x, 45)$ we need to deduct 10 from the x coordinate of $(0, 35)$. Therefore, the coordinates are $(-10, 45)$, and the answer is -10.

18) The correct answer is B. We can see that when $x = 80$, $y = 60$. So, when $x = 160$, $y = 120$. Alternatively, you can determine that the line represents the function: $f(x) = x \times 0.75$. Then substitute 160 for x: $x \times 0.75 = 160 \times 0.75 = 120$

19) The correct answer is C. We simply divide to get the answer: $64 \div 4 = 16$

20) The correct answer is C. Since the figure in this problem is a square, we know that the four sides are equal in length. To find the length of one side, we therefore divide the perimeter by four: $24 \div 4 = 6$

Now we use the Pythagorean Theorem to find the length of line segment AB. In this case AB is the hypotenuse. The hypotenuse length is the square root of $6^2 + 6^2$.

$\sqrt{6^2 + 6^2} =$
$\sqrt{36 + 36} = \sqrt{72}$

So, the answer is $\sqrt{72}$.

21) The correct answer is D. Divide the distance traveled by the time in order to get the speed in miles per hour. Remember that in order to divide by a fraction, you need to invert the fraction, and then multiply.
3.6 miles $\div \, ^3/_4 =$
$3.6 \times \, ^4/_3 =$
$(3.6 \times 4) \div 3 =$
$14.4 \div 3 = 4.8$ miles per hour

22) The correct answer is B. We know that any straight line is 180°. We also know that the sum of the degrees of the three angles of any triangle is 180°. The sum of angles X, Y, and Z = 180. So, the sum of angle X and angle Z equals 180° − 30° = 150°. Remember that in an isosceles triangle, the angles at the base of the triangle are equal. Because this triangle is isosceles, angle X and angle Z are equivalent. So, we can divide the remaining degrees by 2 as follows:
150° ÷ 2 = 75° In other words, angle X and angle Z are each 75°. Then we need to subtract the degree of the angle ∠ZXY from 180° to get the measurement of ∠WXY. 180° − 75° = 105°

23) The correct answer is A. First, we have to calculate the total square footage available.
If there are 4 rooms which are 10 by 10 each, we have this equation:
$4 \times (10 \times 10) = 400$ square feet in total

Now calculate the square footage of the new rooms.
20 × 10 = 200
2 rooms × (10 × 8) = 160
200 + 160 = 360 total square feet for the new rooms

So, the remaining square footage for the bathroom is calculated by taking the total minus the square footage of the new rooms.
400 − 360 = 40 square feet

Since each existing room is 10 feet long, we know that the new bathroom also needs to be 10 feet long in order to fit in. So, the new bathroom is 4 feet × 10 feet.

24) The correct answer is C. The formula for circumference is: $\pi \times 2 \times R$. The center of the circle is on (0, 0) and the top edge of the circle extends to (0, 3), so the radius of the circle is 3. Therefore, the circumference is: $\pi \times 2 \times 3 = 6\pi$

25) The correct answer is B.
The area of circle A is $0.4^2\pi = 0.16\pi$
The area of circle B is $0.2^2\pi = 0.04\pi$
Then subtract: $0.16\pi - 0.04\pi = 0.12\pi$

26) The correct answer is B. A parallelogram is a four-sided figure that has two pairs of parallel sides. The opposite or facing sides of a parallelogram are of equal length and the opposite angles of a parallelogram are of equal measure. You will recall that congruent is another word for equal in measure. So, answer B is correct. A rectangle is a parallelogram with four angles of equal size (all of which are right angles), while a square is a parallelogram with four sides of equal length and four right angles.

27) The correct answer is B. Two angles are supplementary if they add up to 180 degrees.

28) The correct answer is D. To find the mean, add up all of the items in the set and then divide by the number of items in the set. Here we have 7 numbers in the set, so we get our answer as follows: (89 + 65 + 75 + 68 + 82 + 74 + 86) ÷ 7 = 539 ÷ 7 = 77

29) The correct answer is B. At the start of the party, there are 10 red balloons, 7 green balloons, 6 purple balloons, 5 orange balloons, and 11 blue balloons in the bag, so we add all of these up to get our data set at the beginning: 10 + 7 + 6 + 5 + 11 = 39 items in the data set at the beginning. Then a blue balloon and an orange balloon are removed, so we need to reduce the data set for these two items: 39 − 2 = 37. The problem is asking about the probability of a blue balloon. There are 11 blue balloons at the start and one has been removed, so there are 10 blue balloons left. Remember to express the probability as a fraction with the possibility of the outcome (E) in the numerator and the remaining data set (SS) in the denominator: $^{10}/_{37}$

30) The correct answer is D. We have the data set: 1.6, 2.9, 4.5, 2.5, 2.5, 5.1, and 5.4. The mode is the number that occurs most frequently. 2.5 occurs twice, but the other numbers only occur once. So, 2.5 is the mode.

31) The correct answer is B. We don't know the age of the 10th car, so put this in as x to solve:
$(2 + 3 + 4 + 5 + 6 + 7 + 9 + 10 + 12 + x) \div 10 = 6$
$[(2 + 3 + 4 + 5 + 6 + 7 + 9 + 10 + 12 + x) \div 10] \times 10 = 6 \times 10$
$2 + 3 + 4 + 5 + 6 + 7 + 9 + 10 + 12 + x = 60$
$58 + x = 60$
$x = 2$

32) The correct answer is B. Statistical questions usually ask about opinions and behaviors. In addition, statistical questions will have a variety of different answers. On the other hand, questions about measuring time and distance are non-statistical questions because they have only one possible answer. So, the following is a statistical question: "How many residents of the town oppose the tax increase?"

33) The correct answer is A. The range is the difference between the highest number and the lowest number in a data set. If the range increases by 2, then the highest number could go up by 2 or the highest and lowest numbers could increase and decrease by 1 each respectively.

34) The correct answer is B. The dark gray part at the bottom of each bar represents those students who will attend the dance. 45% of the freshman, 30% of the sophomores, 38% of the juniors, and 30% of the seniors will attend. Calculating the average, we get the overall percentage for all four grades: (45 + 30 + 38 + 30) ÷ 4 = 35.75%. 35% is the closest answer to 35.75%, so it best approximates our result.

35) The correct answer is D. In this question, we have an example of a histogram. Histograms are like bar graphs except they show data for groups. To answer these types of questions, be sure that you get the data from the correct group or groups. Here, we need to look at the bars for June 1 at the far right side of the graph. First, find the total amount of accidents on that date. Cars were involved in 30 accidents, vans in 20 accidents, pick-ups in 10 accidents, and SUV's in 5 accidents. So, there were 65 accidents in total (30 + 20 + 10 + 5 = 65). Then divide the number of accidents for pick-ups and vans into the total: 30 ÷ 65 = 46.1538% ≈ 46%

36) The correct answer is B. First of all, add up the amount of faces on the chart: 4 + 3 + 2 + 3 = 12 faces. Each face represents 10 customers, so multiply to get the total number of customers: 12 × 10 = 120 customers in total for all four regions. The salespeople received $540 in total, so we need to divide this by the amount of customers: $540 ÷ 120 customers = $4.50 per customer

37) The correct answer is B. Assign a variable for the age of each boy. Alex = A, Burt = B, and Zander = Z. Alex is twice as old as Burt, so A = 2B. Burt is one year older than three times the age of Zander, so B = 3Z + 1. Then substitute the value of B into the first equation.
A = 2B
A = 2(3Z + 1)
A = 6Z + 2
So, Alex is 2 years older than 6 times the age of Zander.

38) The correct answer is A. We cannot calculate the number of students in the class with belts of particular colors because we do not know the color of the belts the new students. The problem is telling us how many students there are in each group and how many of each group have left. The problem also tells us how many students in total have joined, so we can calculate the new total number of students.

39) The correct answer is D. You have to find the relationship between the number given in each row in the left column and the corresponding number in the right column. "9:50 am to 10:36 am" represents a journey time of 46 minutes. 11:15 to 12:01 is also 46 minutes, and so on. If we go 46 minutes back from 5:51 pm, we arrive at 5:05 pm.

40) The correct answer is D. We don't know how many bottles of strawberry or unscented shampoo were sold. Nor do we know what the store sells normally. So, we cannot calculate the total quantity of shampoo left unsold in the store when it closes on Monday. We can only calculate the quantity of rose-scented shampoo sold since the facts tell us how many bottles of rose-scented shampoo remain in the store at the close of business.

41) The correct answer is C. In order to solve this problem, we would need to multiply the number of gallons of gasoline used per day by the cost of gasoline per gallon by the number of days traveled in order to calculate the total cost. From these required facts, we are lacking the price of gasoline per gallon.

42) The correct answer is B. We know that Paul will have run ten miles when he finishes since he runs 5 miles north, then returns and goes 5 miles south. The question is asking about his running pace or speed. In order to know speed, we need to know the distance traveled and the amount of time it takes to travel the distance. So, we know the distance, but not the time. Accordingly, we would need to know what time he gets back home in order to solve the problem.

43) The correct answer is D. The second fact tells us that if the distance from his house to his destination is less than 5 miles but more than 1 mile, Jose uses his bicycle. If Jose uses his bicycle to go to Manuel's house, then it might be possible that Jose lives 4 miles from Manuel.

44) The correct answer is C. The plumber is going to earn $4,000 for the month. He charges a set fee of $100 per job, and he will do 5 jobs, so we can calculate the total set fees first: $100 set fee per job × 5 jobs = $500 total set fees. Then deduct the set fees from the total for the month in order to determine the total for the hourly pay: $4,000 − $500 = $3,500. He earns $25 per hour, so divide the hourly rate into the total hourly pay in order to determine the number of hours he will work: $3,500 total hourly pay ÷ $25 per hour = 140 hours to work

45) The correct answer is C.

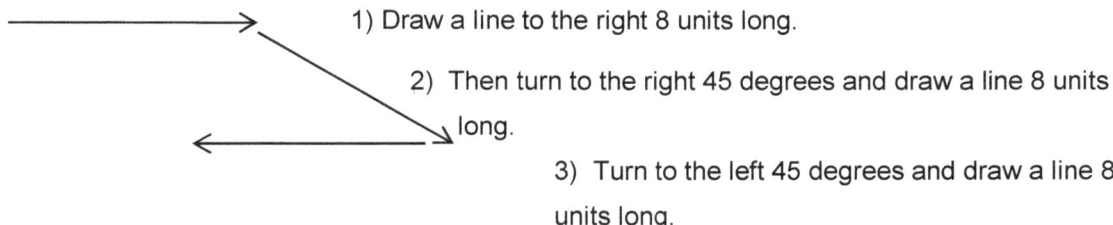

1) Draw a line to the right 8 units long.

2) Then turn to the right 45 degrees and draw a line 8 units long.

3) Turn to the left 45 degrees and draw a line 8 units long.

NES Essential Academic Skills Practice Math Test 4

Number properties and number operations:

1) $82 + 9 \div 3 - 5 = ?$
   A) −40.50
   B) 40.50
   C) 80.00
   D) 85.33

2) $52 + 6 \times 3 - 48 = ?$
   A) 22
   B) 82
   C) 126
   D) 322

3) Convert the following to decimal format: $^3/_{20}$
   A) 0.0015
   B) 0.015
   C) 0.15
   D) 0.66

4) When 1523.48 is divided by 100, which digit of the resulting number is in the tenths place?
   A) 1
   B) 2
   C) 3
   D) 4

5) A group of friends are trying to lose weight. Person A lost $14^3/_4$ pounds. Person B lost $20^1/_5$ pounds. Person C lost 36.35 pounds. What is the total weight loss for the group?
   A) 70.475
   B) 71.05
   C) 71.15
   D) 71.30

6) The total funds, represented by variable F, available for P charity projects is represented by the equation F = $500P + $3,700. If the charity has $40,000 available for projects, what is the greatest number of projects that can be completed?
   A) 72
   B) 73
   C) 74
   D) 79

7) Which of the following is the greatest?
   A) 0.540
   B) 0.054

C) 0.045
D) 0.5045

8) If the value of x is between 0.0007 and 0.0021, which of the following could be x?
A) 0.0012
B) 0.0006
C) 0.0022
D) 0.022

9) Which of the following shows the numbers ordered from greatest to least?
A) $-1/3$, $1/7$, 1, $1/5$
B) $-1/3$, $1/5$, $1/7$, 1
C) $-1/3$, 1, $1/7$, $1/5$
D) 1, $1/5$, $1/7$, $-1/3$

Algebra and graphing:

10) Which of the following is equivalent to the expression $2(x+2)(x-3)$ for all values of x?
A) $2x^2 - 2x - 12$
B) $2x^2 - 10x - 6$
C) $2x^2 + 2x - 12$
D) $2x^2 + 10x - 6$

11) One-half inch on a map represents M miles. Which of the following equations represents $M + 5$ miles on the map?

A) $\dfrac{M+5}{2M}$

B) $\dfrac{0.5M + 2.5}{M}$

C) $\dfrac{2M+5}{M}$

D) $\dfrac{M+5}{2}$

12) The number of visitors a museum had on Tuesday (T) was twice as much as the number of visitors it had on Monday (M). The number of visitors it had on Wednesday (W) was 20% greater than that on Tuesday. Which equation can be used to calculate the total number of visitors to the museum for the three days?
A) W + .20W + 2T + M
B) 2M + T + W
C) M + 1.2T + W
D) 5.4M

13) A construction company is building new homes on a housing development. It has an agreement with the municipality that H number of houses must be built every 30 days. If H number of houses are not built during the 30 day period, the company has to pay a penalty to the municipality of P dollars per house. The penalty is paid per house for the number of houses that fall short of the 30-day target. If A represents the actual number of houses built during the 30-day period, which equation below can be used to calculate the penalty for the 30-day period?
A) $(H - P) \times 30$
B) $(H - A) \times P$
C) $(A - H) \times 30$
D) $(A - H) \times P$

14) If $5 + 5(3\sqrt{x} + 4) = 55$, then $\sqrt{x} = ?$
A) –4
B) –2
C) 2
D) 4

15) State the x and y intercepts that fall on the straight line represented by the following equation:
y = x + 6
A) (–6,0) and (0,6)
B) (0,6) and (0,–6)
C) (6,0) and (0,–6)
D) (0,–6) and (6,0)

16) Consider a two-dimensional linear graph where x = 3 and y = 14. The line crosses the y axis at 5. What is the slope of this line?
A) 2.2
B) 3.0
C) 6.33
D) –2.2

17) Consider the scatterplot below and then choose the best answer from the options that follow.

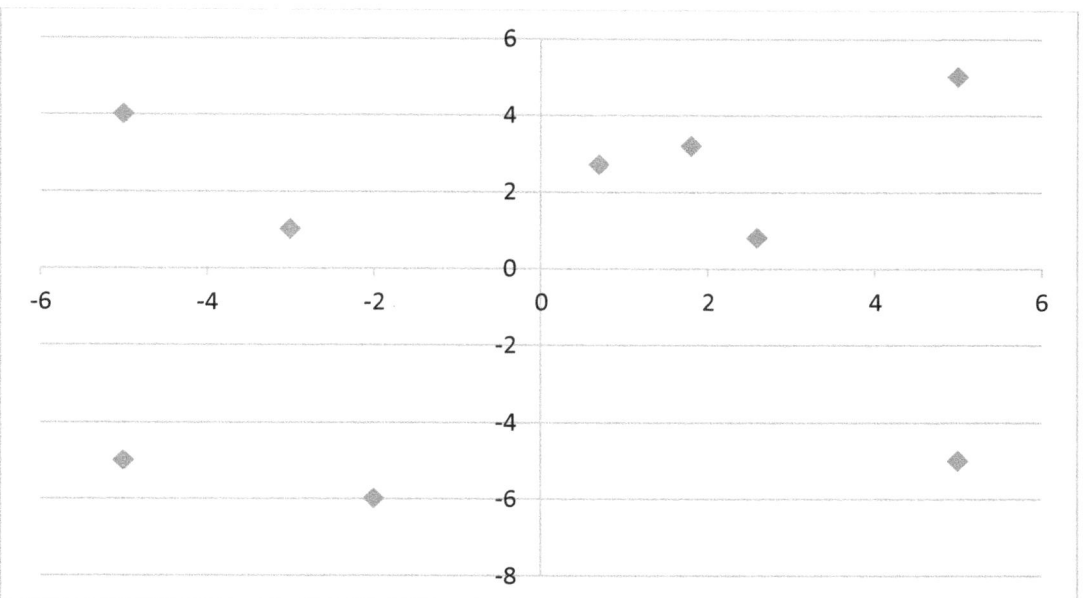

A) The scatterplot suggests a strong positive linear relationship between x and y.
B) The scatterplot suggests a strong negative linear relationship between x and y.
C) The scatterplot suggests a weak positive linear relationship between x and y.
D) The scatterplot suggests that there is no relationship between x and y.

18) A cafeteria serves spaghetti to senior citizens on Fridays. The spaghetti comes prepared in large containers, and each container holds 15 servings of spaghetti. The cafeteria is expecting 82 senior citizens this Friday. What is the least number of containers of spaghetti that the cafeteria will need in order to serve all 82 people?
A) 4
B) 5
C) 6
D) 7

Geometry and measurement:

19) Farmer Brown has a field in which cows craze. He is going to buy fence panels to put up a fence along one side of the field. Each panel is 8 feet 6 inches long. He needs 11 panels to cover the entire side of the field. How long is the field?
A) 60 feet 6 inches
B) 72 feet 8 inches
C) 93 feet 6 inches
D) 102 feet 8 inches

20) A caterpillar travels 10.5 inches in 45 seconds. How far will it travel in 6 minutes?
   A) 45 inches
   B) 63 inches
   C) 64 inches
   D) 84 inches

21) The area of a rectangle is 168 square units. This rectangle contains smaller rectangles that measure 2 square units each. How many of these small rectangles are needed to make up the entire rectangle?
   A) 13
   B) 28
   C) 42
   D) 84

22) Acme Packaging uses string to secure their packages prior to shipment. The string is tied around the entire length and entire width of the package, as shown in the following illustration:

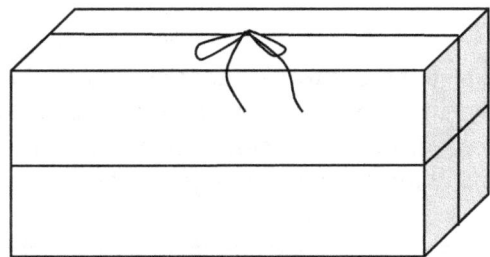

The box is ten inches in height, ten inches in depth, and twenty inches in length. An additional fifteen inches of string is needed to tie a bow on the top of the package. How much string is needed in total in order to tie up the entire package, including making the bow on the top?
   A) 40
   B) 80
   C) 100
   D) 135

23) The triangle in the illustration below is an equilateral triangle. What is the measurement in degrees of angle a?

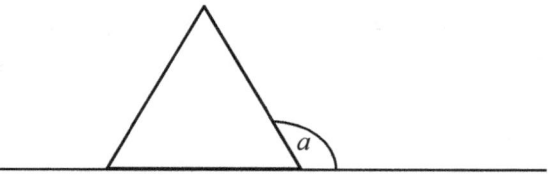

   A) 40
   B) 45
   C) 60
   D) 120

24) The radius (R) of circle A is 5 centimeters. The radius of circle B is 3 centimeters. Which of the following statements is true?

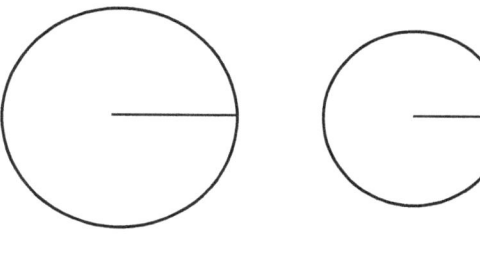

Circle A Circle B

A) The difference between the areas of the circles is 2.
B) The difference between the areas of the circles is 9π.
C) The difference between the circumferences of the circles is 2.
D) The difference between the circumferences of the circles is 4π.

25) A large wheel (L) has a radius of 10 inches. A small wheel (S) has a radius of 6 inches. If the large wheel is going to travel 360 revolutions, how many more revolutions does the small wheel need to make to cover the same distance?
A) 120
B) 240
C) 360
D) 720

26) Consider the vertex of an angle at the center of a circle. The diameter of the circle is 2. If the angle measures 90 degrees, what is the arc length relating to the angle?
A) $\pi/2$
B) $\pi/4$
C) 2π
D) 4π

27) A motorcycle traveled 38.4 miles in $4/5$ of an hour. What was the speed of the motorcycle in miles per hour?
A) 9.6
B) 30.72
C) 48
D) 52

Probability and statistics:

28) The graph below shows the relationship between the number of days of rain per month and the amount of people who exercise outdoors per month. What relationship can be observed?

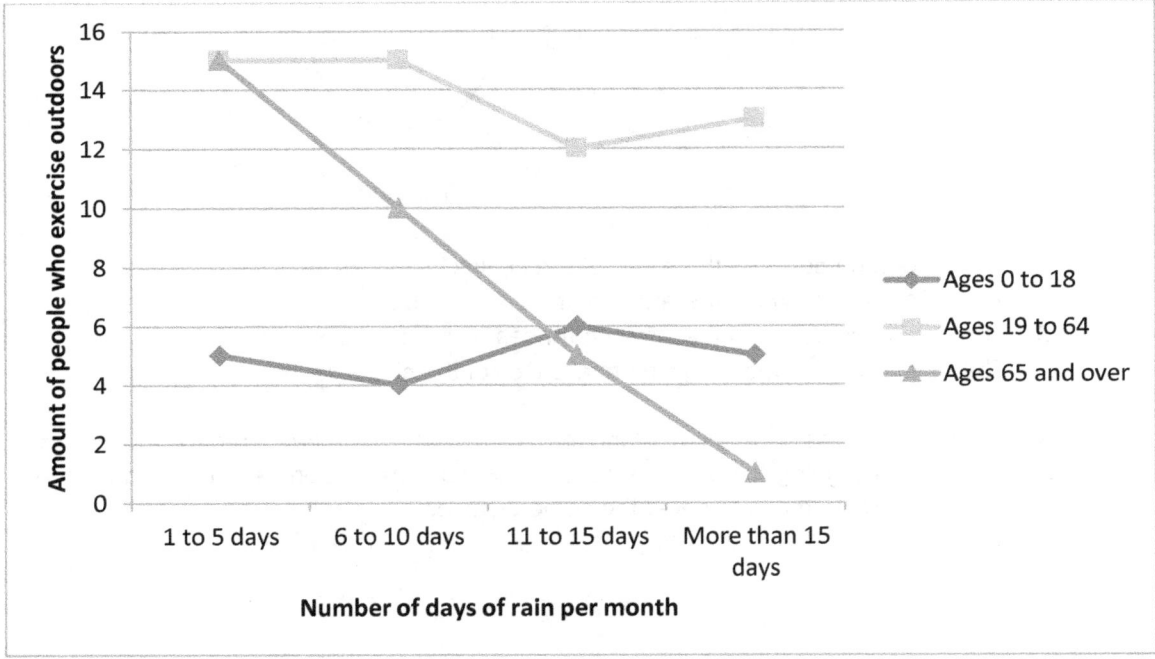

A) Young children are reliant upon an adult in order to exercise outdoors.
B) The exercise habits of working age people seem to fluctuate proportionately to the amount of rainfall.
C) In the 19 to 64 age group, there is a negative relationship between the number of days of rain and the amount of people who exercise outdoors.
D) People aged 65 and over seem less inclined to exercise outdoors when there is more rain.

29) In a group of children, one-half have had a tetanus shot. Of that half, only one-third suffered wounds that would have caused tetanus. In which of the following graphs does the dark gray area represent that third of the group?

A)

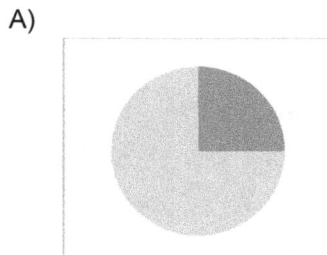

B)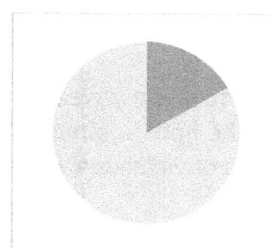

C)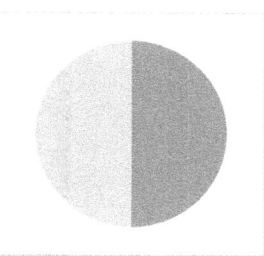

D)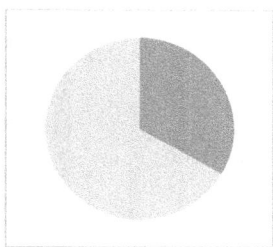

30) The residents of Hendersonville took a census. As part of the census, each resident had to indicate how many relatives they had living within a ten-mile radius of the town. The results of that particular question on the census are represented in the graph below.

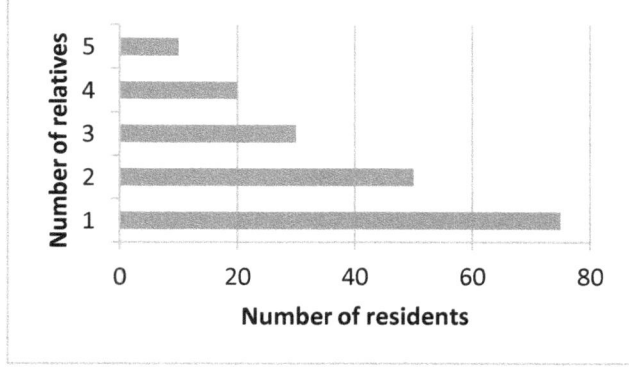

How many residents of Hendersonville had more than 3 relatives living within a ten-mile radius of the town?

A) 10
B) 20

C) 30
D) 155

31) The pictograph below shows the number of traffic violations that occur every week in a certain city. The fine for speeding violations is $50 per violation. The fine for other violations is $20 per violation. The total collected for all three types of violations was $6,000. What is the fine for each parking violation?

Speeding	☆ ☆
Parking	☆
Other violations	☆ ☆ ☆

Each ☆ represents 30 violations.

A) $20
B) $30
C) $40
D) $100

32) Anne has taken a standardized college entrance exam. Use the report of her test scores below to answer the question that follows.

Raw Score Part 1	Raw Score Part 2	Mean	Standard Deviation	Percentile
180	230	205	15	78

Which of the following is a correct interpretation of the score report given above?

A) Ann scored as well as 78% of the test takers.
B) Ann scored as well as 85% of the test takers.
C) 15% of the test takers scored higher than Ann.
D) Ann answered 205 of the questions correctly.

33) Which of these numbers cannot be a probability?
A) 0.02
B) 0
C) 1.002
D) 1

34) An electricity company measures the energy consumption for each home in kilowatt hours (KWH). During July, the homes in one street had the levels of consumption in KWH in the chart show below. What is the mode of the level of energy consumption for this neighborhood for July?

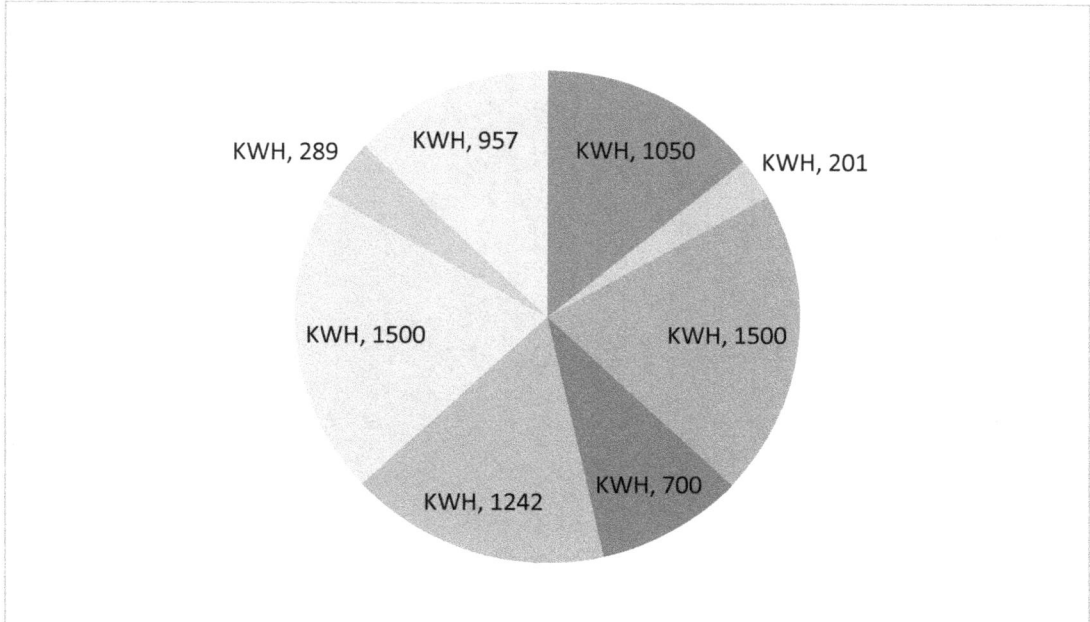

A) 700
B) 957
C) 828.5
D) 1500

35) A die is rolled and a coin is tossed. What is the probability that the die shows an even number and the coin shows tails?

A) $1/2$
B) $1/4$
C) $1/6$
D) $1/12$

36) The blood types of 100 donors are shown in the following chart. If a donation from this group of donors is selected at random, what is the probability that type AB blood will be selected?

Blood Type	Number of donors
A positive	10
A negative	12
B positive	18
B negative	20
O type	25
AB type	15

A) $15/99$
B) $14/100$
C) $3/20$
D) $1/4$

Problem solving, reasoning, and mathematical communication:

37) Use the information below to answer the question that follows.

> Classes will be held every Wednesday morning.
> If there are fewer than 3 children present for a class, the class will be canceled.
> If there is inclement weather, the class will be canceled.

It is Wednesday morning and the class has been canceled. Which one of the following statements is correct?
A) Fewer than three children were present for the class.
B) There was inclement weather.
C) Fewer than three children were present for the class and there was inclement weather.
D) Fewer than three children were present for the class or there was inclement weather.

38) The Abdul family is shopping at a superstore. They buy product A and product B. Product A costs $5 each, and product B costs $8 each. They buy 4 of product A. They also buy a certain quantity of product B. The total value of their purchase is $60. How can we calculate the number of units of product B they bought?
A) Add $5 and $8. Then divide this amount into $60. Then subtract 4 from this result.
B) Subtract $5 from $60. Then divide by $8 and multiply by 4.
C) Multiply 4 by $5. Then deduct this amount from $60. Then divide this result by $8.
D) Multiply 4 by $8. Then deduct this amount from $60. Then divide this result by $5.

39) A scientist would like to investigate the occurrence of seismic events in your state over the past ten years to see whether there is any relationship between the occurrence of seismic events and the amount of rainfall during the same period. Which of the following would be the most suitable method to represent this data?
A) scatterplot
B) pictograph
C) histogram
D) pie chart

40) Classes at a particular university begin at the following times.

| 11:05 am |
| 12:20 pm |
| 1:35 pm |
| 2:50 pm |

If the pattern continues, when does the next class begin?
A) 3:05 pm
B) 3:15 pm
C) 3:45 pm
D) 4:05 pm

41) Four items are going to be placed in a box. The items have the following weights in pounds: 5.14, 4.98, 3.20, 8.78. The box itself weighs 1 pound. What is the best estimate of the total weight of the parcel when the items are placed inside the box?
A) 22
B) 23
C) 28
D) 30

42) The volume of item A is 15 units less than 5 times the volume of item B.
Which of the following equations best expresses the above statement?
A) $A - 15 \times 5 = B$
B) $(A - 15) \times 5 = B$
C) $A = 5B - 15$
D) $A = 15 - 5B$

43) Building A is 15,238 feet high. Building B is 9,427 feet high. Which of the following is the best estimate of the difference between the heights of the two buildings?
A) 5,700
B) 5,800
C) 5,900
D) 6,000

44) A certain county has reported 105,000 cases of cancer per year, which is about 2,000 cases per week.

Which of the following estimates correctly evaluates the reasonableness of the above claim?

A) reasonable, since 100,000 divided by 50 is 2,000
B) unreasonable, since 2,000 times 52 is close to 110,000
C) unreasonable, since 105,000 divided by 52 is 20,000
D) cannot be determined without knowing the number of residents of the county

45) If there is snow on Friday, we will have to cancel the trip.
If the trip is canceled, we will have to file a claim on our travel insurance.
If we file a claim on our travel insurance, the amount of the premium will go up next year.

If the amount of their travel insurance premium went up, which of the following can be inferred?
A) They had to cancel their trip.
B) It snowed on Friday.
C) Both of the above.
D) Neither of the above.

Answer Key – NES Essential Academic Skills Practice Math Test 4

1) C	24) D
2) A	25) B
3) C	26) A
4) B	27) C
5) D	28) D
6) A	29) B
7) A	30) C
8) A	31) C
9) D	32) A
10) A	33) C
11) B	34) D
12) D	35) B
13) B	36) C
14) C	37) D
15) A	38) C
16) B	39) A
17) D	40) D
18) C	41) B
19) C	42) C
20) D	43) B
21) D	44) A
22) D	45) D
23) D	

Answers and Solutions to NES Essential Academic Skills Practice Math Test 4

1) The correct answer is C. Remember that the order of operations is PEMDAS: Parentheses, Exponents, Multiplication, Division, Addition, and Subtraction.
In this problem, there are no operations with parentheses, exponents, or multiplication.
So, do the division first: 82 + (9 ÷ 3) – 5 = 82 + 3 – 5 = 80

2) The correct answer is A. This is another problem on the order of operations.
There are no operations with parentheses or exponents, so do the multiplication first.
52 + (6 × 3) – 48 =
52 + 18 – 48 = 22

3) The correct answer is C. In order to convert a fraction to a decimal, you must divide.

```
      .15
20)3.00
    2.0
    1.00
    1.00
       0
```

4) The correct answer is B. Perform the division, and then check the decimal places of the numbers. Divide as follows: 1523.48 ÷ 100 = 15.2348
Reading our result from left to right: 1 is in the tens place, 5 is in the ones place, 2 is in the tenths place, 3 is in the hundredths place, 4 is in the thousandths place, and 8 is in the ten-thousandths place.

5) The correct answer is D. Represent the mixed numbers as decimal numbers.
Person 1: $14^{3}/_{4}$ = 14.75
Person 2: $20^{1}/_{5}$ = 20.20
Person 3: 36.35

Then add all three amounts together to find the total.
14.75 + 20.20 + 36.35 = 71.30

6) The correct answer is A. The equation is: F = $500P + $3,700. We are told that the total funds are $40,000 so put that in the equation to solve the problem.
$40,000 = $500P + $3,700
$40,000 – $3,700 = $500P
$36,300 = $500P
$36,300 ÷ 500 = $500 ÷ 500P
$36,300 ÷ 500 = 72.6
Since a fraction of a project cannot be undertaken, the greatest number of projects is 72.

7) The correct answer is A. For problems with decimals, line the figures up in a column and add zeroes to fill in the column as shown.

0.5400
0.0540
0.0450
0.5045

If you struggle with decimals, you can remove the decimal points and the zeroes before the other integers in order to see the answer more clearly.

~~0.~~5400
~~0.0~~540
~~0.0~~450
~~0.~~5045

When we have removed the zeroes in front of the other numbers, we can see that the largest number is the first one, which is 0.54.

8) The correct answer is A. Like the solution above, put in zeroes and line up the decimal points when you compare the numbers.

    0.0007
A) 0.0012
B) 0.0006
C) 0.0022
D) 0.0220
    0.0021

Answer choice B is less than 0.0007, and answer choices C and D are greater than 0.0021. Answer choice A (0.0012) is between 0.0007 and 0.0021, so it is the correct answer.

9) The correct answer is D. Remember these principles: (a) Positive numbers are greater than negative numbers; (b) When two fractions have the same numerator, the fraction with the smaller number in the denominator is the larger fraction. Accordingly, 1 is greater than $1/5$; $1/5$ is greater than $1/7$, and $1/7$ is greater than $-1/3$.

10) The correct answer is A. You should use the FOIL method in this problem. Be very careful with the negative numbers when doing the multiplication.
$2(x + 2)(x - 3) =$
$2[(x \times x) + (x \times -3) + (2 \times x) + (2 \times -3)] =$
$2(x^2 + -3x + 2x + -6) =$
$2(x^2 - 3x + 2x - 6) =$
$2(x^2 - x - 6)$

Then multiply each term by the 2 at the front of the parentheses.
$2(x^2 - x - 6) =$
$2x^2 - 2x - 12$

11) The correct answer is B. The ratio of 0.5 inch for M miles can be represented mathematically as $\frac{0.5}{M}$. The ratio for $M + 5$ is not known, so we can represent the unknown as x: $\frac{x}{M+5}$. Finally, use cross multiplication to solve the problem:

$$\frac{0.5}{M} = \frac{x}{M+5}$$

$$0.5 \times (M+5) = Mx$$

Then divide by M to isolate x and solve the problem.

$$[0.5 \times (M+5)] \div M = Mx \div M$$

$$\frac{0.5M + 2.5}{M} = x$$

12) The correct answer is D. Set up each part of the problem as an equation. The museum had twice as many visitors on Tuesday (T) as on Monday (M), so T = 2M. The number of visitors on Wednesday exceeded that of Tuesday by 20%, so W = 1.20 × T. Then express T in terms of M for Wednesday's visitors: W = 1.20 × T = 1.20 × 2M = 2.40M. Finally, add the amounts together for all three days: M + 2M + 2.40M = 5.4M

13) The correct answer is B. First, we need to calculate the shortage in the amount of houses actually built. If H represents the amount of houses that should be built and A represents the actual number of houses built, then the shortage is calculated as: $H - A$. The company has to pay P dollars per house for the shortage, so we calculate the total penalty by multiplying the shortage by the penalty per house: $(H - A) \times P$

14) The correct answer is C. First, deal with the integers that are outside the parentheses.

$5 + 5(3\sqrt{x} + 4) = 55$

$5 + 15\sqrt{x} + 20 = 55$

$25 + 15\sqrt{x} = 55$

$25 - 25 + 15\sqrt{x} = 55 - 25$

$15\sqrt{x} = 30$

Then divide in order to isolate $\sqrt{x}$.

$15\sqrt{x} = 30$

$(15\sqrt{x}) \div 15 = 30 \div 15$

$\sqrt{x} = 2$

15) The correct answer is A. The y intercept is where the line crosses the y axis, so x = 0 for the y intercept.

Begin by substituting 0 for x.
y = x + 6
y = 0 + 6
y = 6

Therefore, the coordinates (0, 6) represent the y intercept.
On the other hand, the x intercept is where the line crosses the x axis, so y = 0 for the x intercept. Now substitute 0 for y.

$y = x + 6$
$0 = x + 6$
$0 - 6 = x + 6 - 6$
$-6 = x$
So, the coordinates (–6, 0) represent the x intercept.

16) The correct answer is B. When you are provided with a set of coordinates and the y intercept, you need the slope-intercept formula in order to calculate the slope of a line.
In the slope-intercept formula, m is the slope and b is the y intercept, which is the point where the line crosses the y axis. Now solve for the numbers given in the problem.
$y = mx + b$
$14 = m3 + 5$
$14 - 5 = m3 + 5 - 5$
$9 = m3$
$9 \div 3 = m$
$3 = m$

17) The correct answer is D. When looking at scatterplots, try to see if the dots are roughly grouped into any kind of pattern or line. If so, positive or negative relationships may be represented. Here, however, the dots are located at what appear to be random places on all four quadrants of the graph. So, the scatterplot suggests that there is no relationship between x and y.

18) The correct answer is C. Divide and then round up: 82 people in total ÷ 15 people served per container = 5.467 containers. We need to round up to 6 since we cannot purchase a fractional part of a container.

19) The correct answer is C. Each panel is 8 feet 6 inches long, and he needs 11 panels to cover the entire side of the field. So, we need to multiply 8 feet 6 inches by 11, and then simplify the result. Step 1: 8 feet × 11 = 88 feet; Step 2: 6 inches × 11 = 66 inches; Step 3: There are 12 inches in a foot, so we need to determine how many feet and inches there are in 66 inches. 66 inches ÷ 12 = 5 feet 6 inches; Step 4: Now add the two results together. 88 feet + 5 feet 6 inches = 93 feet 6 inches

20) The correct answer is D. The question is asking us about a time duration of 6 minutes, so we need to calculate the amount of seconds in 6 minutes: 6 minutes × 60 seconds per minute = 360 seconds in total. Then divide the total time by the amount of time needed to make one journey: 360 seconds ÷ 45 seconds per journey = 8 journeys. Finally, multiply the number of journeys by the amount of inches per journey in order to get the total inches: 10.5 inches for 1 journey × 8 journeys = 84 inches in total

21) The correct answer is D. A rectangle consisting of 2 square units will look like the following illustration:

So, we divide the total number of squares in the larger rectangle by 2: 168 ÷ 2 = 84

22) The correct answer is D. The string that goes around the front, back, and sides of the package is calculated as follows: 20 + 10 + 20 + 10 = 60. The string that goes around the top, bottom, and sides of the package is calculated in the same way since the top and bottom are

equal in length to the front and back: 20 + 10 + 20 + 10 = 60. So, 120 inches of string is needed so far. Then, we need 15 extra inches for the bow: 120 + 15 = 135

23) The correct answer is D. An equilateral triangle has three equal sides and three equal angles. Since all 3 angles in any triangle need to add up to 180 degrees, each angle of an equilateral triangle is 60 degrees (180 ÷ 3 = 60). Angles that lie along the same side of a straight line must add up to 180. So, we calculate angle a as follows: 180 − 60 = 120

24) The correct answers is D. The formula for the area of a circle is: πR^2. The area of circle A is $\pi \times 5^2 = 25\pi$ and the area of circle B is $\pi \times 3^2 = 9\pi$. So, the difference between the areas is 16π. The formula for circumference is: $\pi 2R$. The circumference of circle A is $\pi \times 2 \times 5 = 10\pi$ and the circumference for circle B is $\pi \times 2 \times 3 = 6\pi$. The difference in the circumferences is 4π. So, answer D is correct.

25) The correct answer is B. Circumference is $2\pi R$, so the circumference of the large wheel is 20π and the circumference of the smaller wheel is 12π. If the large wheel travels 360 revolutions, it travels a distance of: $20\pi \times 360 = 7200\pi$. To determine the number of revolutions the small wheel needs to make to go the same distance, we divide the distance by the circumference of the smaller wheel: $7200\pi \div 12\pi = 600$. Finally, calculate the difference in the number of revolutions: $600 - 360 = 240$

26) The correct answer is A. For questions like this on arcs, you should first find the circumference of the circle. The diameter of the circle is 2, so the circumference is 2π. There are 360 degrees in a circle and the question is asking us about a 90 degree angle, so the arc length relates to one-fourth of the circumference: 90 ÷ 360 = $1/4$. So, we need to take one-fourth of the circumference to get the answer: $2\pi \times 1/4 = 2\pi/4 = \pi/2$

27) The correct answer is C. Divide by the fractional hour in order to determine the speed for an entire hour: 38.4 miles ÷ $4/5$ of an hour = 38.4 × $5/4$ = (38 × 5) ÷ 4 = 48 mph

28) The correct answer is D. The most striking relationship on the graph is the line for ages 65 and over, which clearly shows a negative relationship between exercising outdoors and the number of days of rain per month. You will recall that a negative relationship exists when an increase in one variable causes a decrease in the other variable. So, we can conclude that people aged 65 and over seem less inclined to exercise outdoors when there is more rain.

29) The correct answer is B. The question is asking us to calculate one third of one half. So, we multiply to get our answer: $1/2 \times 1/3 = (1 \times 1)/(2 \times 3) = 1/6$

30) The correct answer is C. The question is asking us how many residents have more than 3 relatives nearby, so we need to add the bars for 4 and 5 relatives from the chart. 20 residents have 4 relatives nearby and 10 residents have 5 relatives nearby, so 30 residents (20 + 10 = 30) have more than 3 relatives nearby.

31) The correct answer is C. There are 2 stars for speeding, and each star equals 30 violations, so there were 60 speeding violations in total. The fine for speeding is $50 per violation, so the total amount collected for speeding violations was: 60 speeding violations × $50 per violation = $3000. There are three stars for other violations, which is equal to 90 violations (3 × 30 = 90).

Other violations are $20 each, so the total for other violations is: 90 × $20 = $1800. Next, we need to deduct these two amounts from the total collections of $6,000 in order to find out how much was collected for parking violations: $6000 − $3000 − $1800 = $1200 in total for parking violations. There is one star for parking violations, so there were 30 parking violations. We divide to get the answer: $1200 income for parking violations ÷ 30 parking violations = $40 each

32) The correct answer is A. The raw score represents the number of questions that were answered correctly. The mean is the average of the first two raw scores. In other words, we can calculate the mean like this: (180 + 230) ÷ 2 = 205

Standard deviation measures the variation from the mean or average. The percentile rank of a score is the percentage of test-takers that scored the same or lower than the student in question. For instance, a percentile score of 60 means that 60% of the test-takers scored the same or lower than a particular student. In our question, Anne's score were in the 78th percentile, so Ann scored as well as 78% of the test takers.

33) The correct answer is C. Probability will be 1 for a 100% probability, 0 for something that has no change of occurring, or a positive number less than 1 for all other probabilities. Probability can be expressed as a decimal or a fraction. Probability cannot be a negative number or a number greater than 1.

34) The correct answer is D. The mode is the number in the set that occurs most frequently. Our data set is: 1050, 201, 1500, 700, 1242, 1500, 289, 957. The number 1500 is the only number that occurs more than once, so it is the mode.

35) The correct answer is B. The data set can be expressed as follows:

(1,H), (2,H), (3,H), (4,H), (5,H), (6,H), (1,T), (2,T), (3,T), (4,T), (5,T), (6,T)

Counting the items in the above set, we can see that there are 12 items in total. The desired outcome is that the die shows an even number and the coin shows a tails. The possible outcomes are: {(2,T),(4,T),(6,T)}

So, the probability is: $3/12 = 1/4$

36) The correct answer is C. The total for the data set is: 10 + 12 + 18 + 20 + 25 + 15 = 100. There are 15 donors with type AB blood, so the probability is $15/100 = 3/20$

37) The correct answer is D. The second fact tells us that if there are fewer than 3 children present for a class, the class will be canceled. The third fact tells us that if there is inclement weather, the class will also be canceled.

38) The correct answer is C. Multiply 4 by $5. Then deduct this amount from $60. Then divide this result by $8.

Your equation is: (B × $8) = $60 − (4 × $5)
This is simplified to: (B × $8) = $40
Then divide to get the number of units: B = $40 ÷ $8

39) The correct answer is A. As stated in the tips in practice test 1, scatterplots are useful when we are trying to determine whether a relationship exists between two events. In this question, we are examining whether there is a relationship between seismic events and rainfall.

40) The correct answer is D. Each class is an hour and fifteen minutes long. Adding an hour and fifteen minutes to the last time, which is 2:50 pm, give us an answer of 4:05 pm.

41) The correct answer is B. In order to make the estimate, round each item up or down to the nearest pound. The weights in the problem were 5.14, 4.98, 3.20, 8.78. We round these to 5, 5, 3, and 9. Then add these together and add 1 more pound for the box: 5 + 5 + 3 + 9 + 1 = 23

42) The correct answer is C. The problem states that the volume of item A is 15 units less than 5 times the volume of item B. So set up your equation based on each part of the question.

5 times the volume of item B = 5B

The volume of A is 15 less than 5 times the volume of item B, so A = 5B – 15

43) The correct answer is B. We have to subtract to find the difference in height between the two buildings. First of all, round each number up or down. Looking at the answer choices, we can see that we need to round to the nearest increment of 100. 15,238 is rounded down to 15,200 and 9,427 is rounded down to 9,400. Now subtract to get your answer: 15,200 – 9,400 = 5,800

44) The correct answer is A. Remember to do a calculation based on each answer choice. A is correct because 100,000 ÷ 50 = 2,000.

45) The correct answer is D. There may be other causes for the result apart from the snow. For instance, inflation will also cause prices to increase.

NES Essential Academic Skills Practice Math Test 5

Number properties and number relationships:

1) $6\tfrac{3}{4} - 2\tfrac{1}{2} = ?$
   A) $4\tfrac{1}{4}$
   B) $4\tfrac{3}{8}$
   C) $4\tfrac{5}{8}$
   D) $4\tfrac{6}{8}$

2) $9 \times 6 + 42 \div 6 = ?$
   A) 8
   B) 16
   C) 27
   D) 61

3) Find the value of x that solves the following proportion: $\tfrac{9}{6} = \tfrac{x}{10}$
   A) 1.5
   B) 15
   C) .67
   D) 67

4) $\tfrac{1}{8} \div \tfrac{4}{3} = ?$
   A) $\tfrac{1}{6}$
   B) $\tfrac{32}{3}$
   C) $\tfrac{3}{24}$
   D) $\tfrac{3}{32}$

5) Convert the following fraction to decimal format: $\tfrac{5}{50}$
   A) 0.0010
   B) 0.0100
   C) 0.1000
   D) 0.0500

6) $\tfrac{1}{6} + (\tfrac{1}{2} \div \tfrac{3}{8}) - (\tfrac{1}{3} \times \tfrac{3}{2}) = ?$
   A) $\tfrac{23}{6}$
   B) 1
   C) 2
   D) $\tfrac{1}{10}$

7) Mary needs to get $650 in donations. So far, she has obtained 80% of the money she needs. How much money does she still need?
   A) $8.19
   B) $13.00
   C) $32.50
   D) $130.00

8) $(6y)^0 = ?$
   A) 6y
   B) 6
   C) 1
   D) 0

9) $(-5)^{-2} = ?$
   A) −25
   B) −1/25
   C) 1/25
   D) 25

Algebra and graphing:

10) Use the chart below to answer the question that follows.

X	Y
2	4
4	16
6	
8	64
10	100

The chart above shows the mathematical relationship between X and Y. What value of Y is missing from the chart?
A) 24
B) 30
C) 32
D) 36

11) For all positive integers x and y, x − 6 < 0 and y < x + 12, then y < ?
A) 6
B) 12
C) 18
D) 24

12) Which of the following is the graph of the solution of 2 + y < −8?

A)
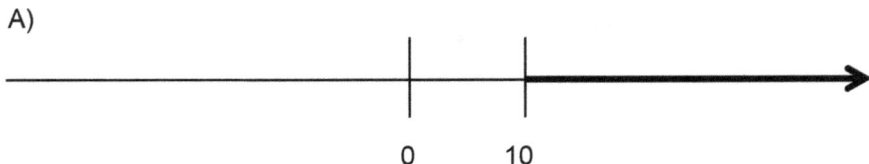

B)

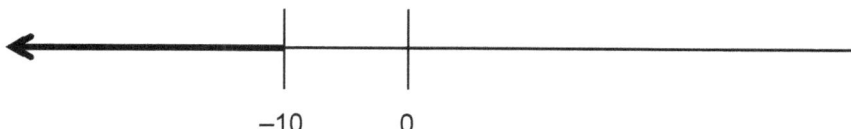

C)

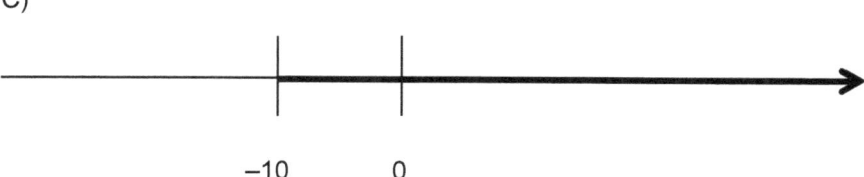

D)

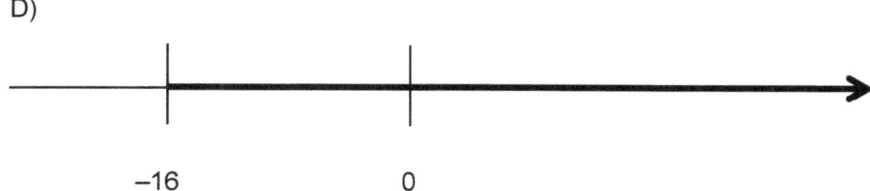

13) The graph of a linear equation is shown below. Which one of the tables of values best represents the points on the graph?

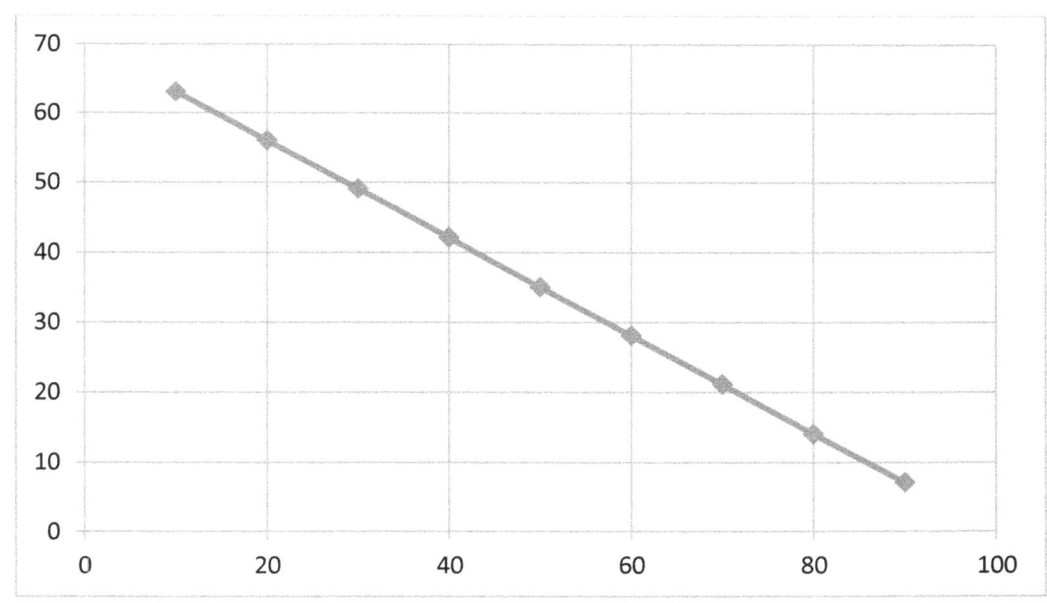

A)

x	y
5	65
10	64
15	63
20	62

B)

x	y
5	68
15	60
25	52
35	54

C)

x	y
10	63
20	56
30	49
40	42

D)

x	y
10	68
20	60
30	52
40	44

14) Which of the following equations is equivalent to $\frac{x}{5} \div \frac{9}{y}$?

A) $\frac{xy}{45}$

B) $\frac{9x}{5y}$

C) $\frac{1}{9} \times \frac{x}{5y}$

D) $\frac{1}{5} \times \frac{9}{5y}$

15) A factory that makes microchips produces 20 times as many functioning chips as defective chips. If the factory produced 11,235 chips in total last week, how many of them were defective?
A) 535
B) 561
C) 1,070
D) 10,700

16) The price of a wool coat is reduced 12.5% at the end of the winter. If the original price of the coat was $120, what will the price be after the reduction?
A) $108.00
B) $107.50
C) $105.70
D) $105.00

17) If $\frac{1}{5}x + 3 = 5$, then $x = ?$

A) $\frac{8}{5}$

B) $-\frac{8}{5}$

C) 8

D) 10

18) Which of the following is equivalent to the expression $36 - 2x$ for all values of x?
A) $6 + 2(15 - x)$
B) $6(6 - 2x)$
C) $39 - (3 - 2x)$
D) $8(5 - 2x) - 4$

Geometry and measurement:

19) Consider two concentric circles with radii of $R_1 = 1$ and $R_2 = 2.4$ as shown in the illustration below. Line L forms the diameter of the circles. What is the area of the lined part of the illustration?

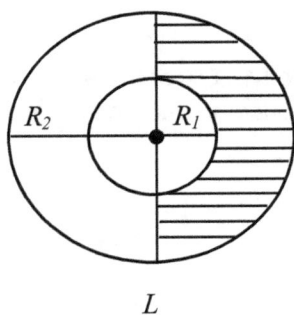

A) 0.7π
B) 1.4π
C) 2π
D) 2.38π

20) Liz wants to put new vinyl flooring in her kitchen. She will buy the flooring in square pieces that measure 1 square foot each. The entire room is 8 feet by 12 feet. The cupboards are two feet deep from front to back. Flooring will not be put under the cupboards. A diagram of her kitchen is provided.

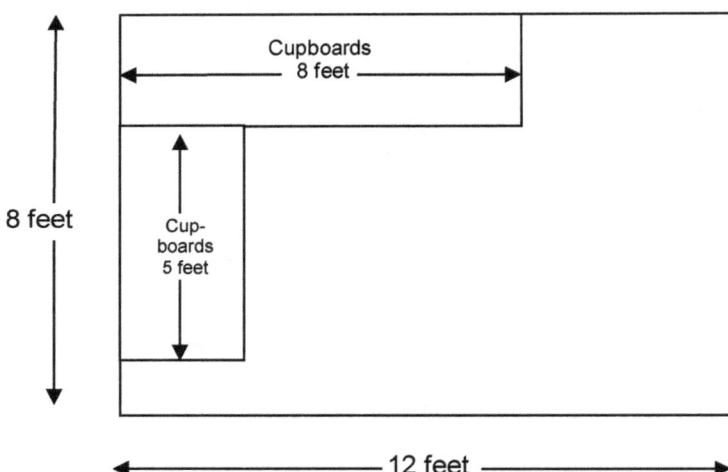

How many pieces of vinyl will Liz need to cover her floor?
A) 120
B) 96
C) 70
D) 84

21) A vegetable grower has a rectangular pen in which she keeps her produce. She has decided to divide the pen into two parts. To divide the pen, she will erect a fence diagonally from the two corners, as shown in the diagram below. How long in yards is the diagonal fence?

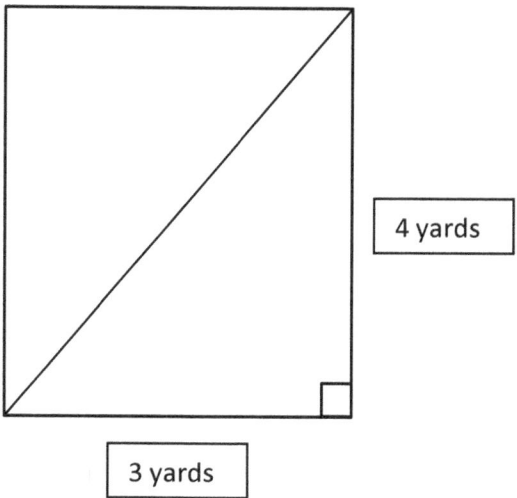

A) 4
B) 5
C) 5.5
D) 6

22) The diagram below shows a figure made from a semicircle, a rectangle, and an equilateral triangle. The rectangle has a length of 18 inches and a width of 10 inches. What is the perimeter of the figure?

A) 56 inches + 5π inches
B) 56 inches + 10π inches
C) 56 inches + 12.5π inches
D) 56 inches + 25π inches

23) In the figure below, the circle centered at B is internally tangent to the circle centered at A. The length of line segment AB, which represents the radius of circle A, is 3 units and the smaller circle passes through the center of the larger circle. If the area of the smaller circle is removed from the larger circle, what is the remaining area of the larger circle?

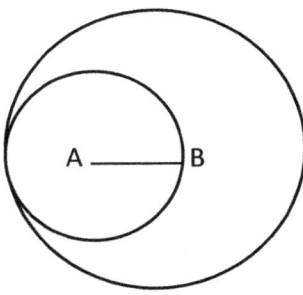

A) 3π
B) 6π
C) 9π
D) 27π

24) The perimeter of a rectangle is 48 meters. If the length were doubled and the width were increased by 5 meters, the perimeter would be 92 meters. What are the length and width of the original rectangle?
A) width = 7, length = 17
B) width = 17, length = 7
C) width = 34, length = 14
D) width = 24, length = 46

25) Use the diagram below to answer the question that follows.

Each square in the diagram above is one foot wide and one foot long. The gray area of the diagram represents the layout of New Town's water reservoir. What is the perimeter in feet of the reservoir?
A) 16 feet
B) 17 feet
C) 18 feet
D) 20 feet

26) Captain Smith needs to purchase rope for his fleet of yachts. He owns 26 yachts and needs 6 feet 10 inches of rope for each one. How much rope does he need in total?
A) 152 feet
B) 177 feet 8 inches
C) 257 feet 8 inches
D) 260 feet

27) A photograph measures 4 inches by 6 inches. Tom wants to make a wooden frame for the photo. He will cut the wood into 4 pieces, but he needs an extra inch in length on each piece of wood to finish off the corners. What total length of wood will he need in order to complete the project?
A) 10 inches
B) 12 inches
C) 16 inches
D) 24 inches

Probability and statistics:

28) Which of the following is a statistical question?
A) How long is that piece of string?
B) How many customers go to that coffee shop on Saturdays?
C) What time is algebra class?
D) What are the perimeter measurements of the high school parking lot?

29) The state highway department wants to find out how the residents of Buford feel about a new road being constructed around their town. Which one of the following methods will result in the most valid information about the opinions of the residents at the town?
A) To question drivers of a random selection of vehicles traveling through Buford
B) To poll drivers at random as they exit the interstate highway near Buford
C) To ask car owners living in Buford to participate in a survey
D) To select participants for a survey from a list of all of the citizens living in Buford

30) A deck of cards contains 13 hearts, 13 diamonds, 13 clubs, and 13 spades. Cards are selected from the deck at random. Once selected, the cards are discarded and are not placed back into the deck. Two spades, one heart, and a club are drawn from the deck. What is the probability that the next card drawn from the deck will be a heart?
A) $1/13$
B) $1/12$

C) 13/52
D) 1/4

31) What is the median of the numbers in the following list?:
2.5, 9.4, 3.1, 1.7, 3.2, 8.2, 4.5, 6.4, 7.8
A) 3.2
B) 4.5
C) 5.2
D) 6.4

32) An airplane flew at a constant speed, traveling 780 miles in 2 hours. The graph below shows the total miles the airplane traveled in 20 minute intervals. According to the graph, how many miles did the plane travel in the last 40 minutes of its journey?

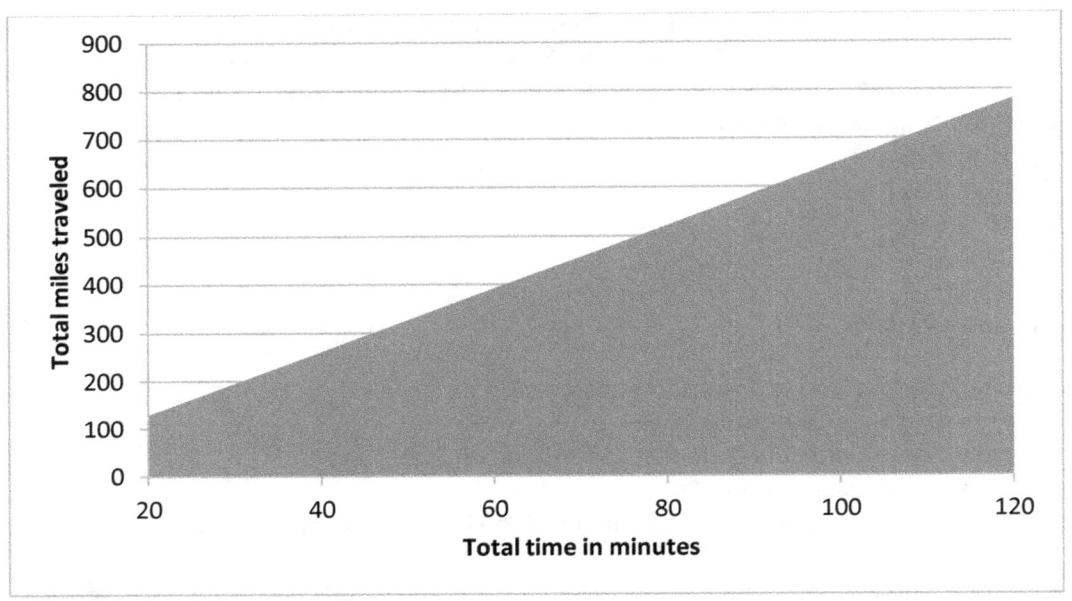

A) 120
B) 180
C) 200
D) 260

33) A family is planning an annual picnic in Arizona. Rain is forecast for 45 days of the year, but when rain is forecast, the prediction is correct only 90% of the time. What is the probability that it will rain on the day of the picnic? Note that it is not a leap year.
A) 2.2222%
B) 11.0959%
C) 12.3288%
D) 45%

34) In an athletic competition, the maximum possible amount of points was 25 points per participant. The scores for 15 different participants are displayed in the graph below. What was the median score for the 15 participants?

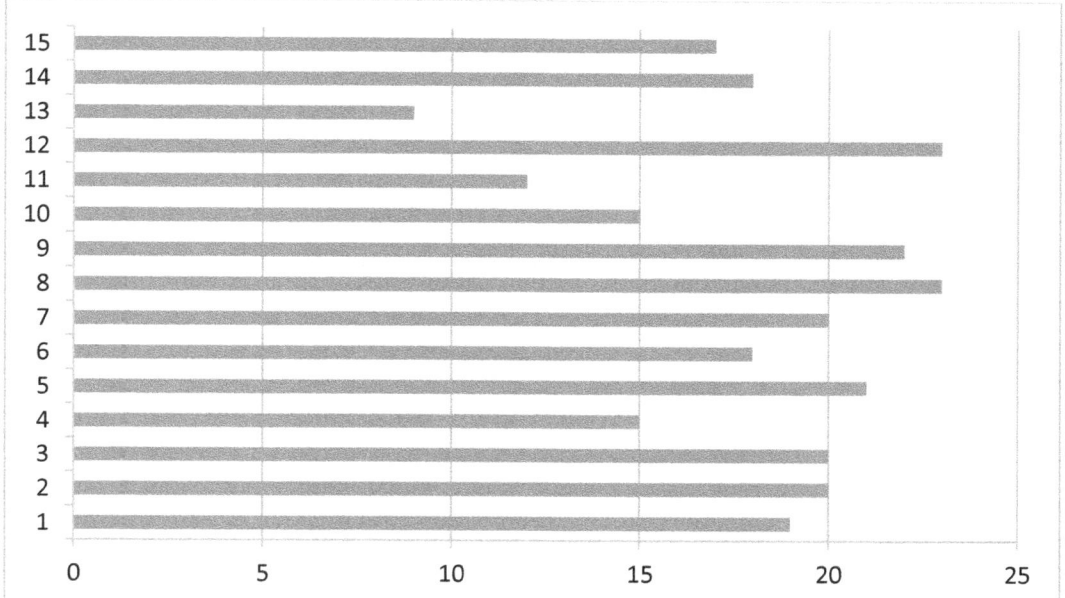

A) 8
B) 15
C) 16
D) 19

35) Suki rolls a fair pair of six-sided dice. Each die has values from 1 to 6. She rolls an even number on her first roll. What is the probability that she will roll an odd number on her next roll?

A) $1/2$
B) $1/6$
C) $2/6$
D) $6/11$

36) Which of the following methods of data representation would be best to represent the percentage that each class is to the total enrollment for a particular high school at the start of this academic year?
A) pie chart
B) scatterplot
C) bar graph
D) pictograph

Problem solving, reasoning, and mathematical communication:

37) Use the information below to answer the question that follows.

> The supermarket is 12 miles away from the gas station.
> Tom's house is 18 miles away from the gas station.

Based on the information given above, which one of the following statements is correct?
A) Tom's house is 6 miles from the supermarket.
B) Tom's house is 12 miles from the supermarket.
C) Tom's house is no more than 18 miles from the supermarket.
D) Tom's house is no more than 30 miles from the supermarket.

38) A is 3 times B, and B is 3 more than 6 times C. Which of the following describes the relationship between A and C?
A) A is 9 more than 18 times C.
B) A is 3 more than 3 times C.
C) A is 3 more than 18 times C.
D) A is 6 more than 3 times C.

39) Toby is going to buy a car. The total purchase price of the car is represented by variable C. He will pay D dollars immediately, and then he will make equal payments (P) each month for a certain number of months (M). Which instructions below can be used to calculate the amount of his monthly payment (P)?
A) Subtract D from C. Then divide this amount by M.
B) Divide C by M. Then subtract D from this amount.
C) Subtract D from C. Then divide this result into M.
D) Divide C by M. Then subtract this result from D.

40) During each flight, a flight attendant must count the number of passengers on board the aircraft. The morning flight had 52 passengers more than the evening flight, and there were 540 passengers in total on the two flights that day. How many passengers were there on the evening flight?
A) 244
B) 296
C) 488
D) 540

41) A dance academy had 300 students at the beginning of January. It lost 5% of its students during the month. However, 15 new students joined the academy on the last day of the month. If this pattern continues for the next two months, how many students will there be at the academy at the end of March?
A) 285
B) 300
C) 310
D) 315

42) A customer buys five items at the following prices: $1.99, $20.05, $98.75, $0.35, $103.62
What is the best estimate of the total value of these five items?
A) $180
B) $200
C) $225
D) $250

43) If the pattern continues, what is the next item in the following list?
$$\frac{1}{4}, \frac{5}{8}, \frac{9}{12}, \frac{13}{16}$$
A) $\frac{14}{20}$
B) $\frac{17}{20}$
C) $\frac{18}{19}$
D) $\frac{21}{19}$

44) 5 more than 4 times the number x is equal to the number y.
Which statement below represents this equation?
A) 5x + 4 = y
B) 4x + 5 = y
C) 4(x + 5) = y
D) 4(y + 5) = x

45) You are inviting Tom, Maria, Paul, Sue, and Jeri to your house for a meal. Tom won't sit next to Paul, and Paul won't sit next to Maria. Jeri has to sit next to Sue. You have a circular table for you and your five guests. Which of the following is a suitable seating plan?
A) Tom, Maria, Paul, Sue, Jeri, You
B) Tom, Paul, Maria, Sue, Jeri, You
C) Paul, Sue, Jeri, Tom, Maria, You
D) Tom, Paul, Sue, Maria, Jeri, You

Answer Key – NES Essential Academic Skills Practice Math Test 5

1) A
2) D
3) B
4) D
5) C
6) B
7) D
8) C
9) C
10) D
11) C
12) B
13) C
14) A
15) A
16) D
17) D
18) A
19) D
20) C
21) B
22) A
23) D

24) A
25) C
26) B
27) D
28) B
29) D
30) D
31) B
32) D
33) B
34) D
35) A
36) A
37) D
38) A
39) A
40) A
41) B
42) C
43) B
44) B
45) C

Answers and Solutions to NES Essential Academic Skills Practice Math Test 5

1) The correct answer is A. Questions like this test your knowledge of mixed numbers. If the fraction on the first mixed number is greater than the fraction on the second mixed number, you can subtract the whole numbers and the fractions separately. Remember to use the lowest common denominator on the fractions. First, subtract whole numbers: 6 − 2 = 4

Then subtract fractions.
$3/4 − 1/2 =$
$3/4 − 2/4 =$
$1/4$

Now put them together for the result.
$4 \frac{1}{4}$

Alternatively, do the operations as follows:
$6\frac{3}{4} - 2\frac{1}{2} =$
$6\frac{3}{4} - [2 + (\frac{1}{2} \times \frac{2}{2})] =$
$6\frac{3}{4} - 2\frac{2}{4} = 4\frac{1}{4}$

2) The correct answer is D. Use PEMDAS: Parentheses, Exponents, Multiplication, Division, Addition, and Subtraction. Do the division and multiplication first, before adding or subtracting:
9 × 6 + 42 ÷ 6 = (9 × 6) + (42 ÷ 6) = 54 + 7 = 61

3) The correct answer is B. You can simplify the first fraction because both the numerator and denominator are divisible by 3: $9/6 \div 3/3 = 3/2$
Then divide the denominator of the second fraction by the denominator 2 of the simplified fraction $3/2$ from above: 10 ÷ 2 = 5
Now, multiply this number by the numerator of the first fraction to get your result: 5 × 3 = 15
You can check your answer as follows:
$9/6 = 15/10$
$9/6 \div 3/3 = 3/2$
$15/10 \div 5/5 = 3/2$

4) The correct answer is D. When you are asked to divide fractions, remember that you need to invert the second fraction. Then you multiply this inverted fraction by the first fraction given in the problem. $4/3$ inverted is $3/4$. Then multiply the numerators and the denominators together to get the new fraction.

$$\frac{1}{8} \div \frac{4}{3} =$$
$$\frac{1}{8} \times \frac{3}{4} = \frac{3}{32}$$

5) The correct answer is C. To represent a fraction as a decimal, you need to divide. So, you will need to do long division to determine the answer.

```
       .10
50)5.00
      5.00
         0
```

6) The correct answer is B. Invert the second fraction and then multiply the fractions. For problems like this, deal with the parts of the equation in the parentheses first.

$$\frac{1}{6} + \left(\frac{1}{2} \div \frac{3}{8}\right) - \left(\frac{1}{3} \times \frac{3}{2}\right) =$$

$$\frac{1}{6} + \left(\frac{1}{2} \times \frac{8}{3}\right) - \left(\frac{1}{3} \times \frac{3}{2}\right) =$$

$$\frac{1}{6} + \frac{8}{6} - \frac{3}{6}$$

After you have done the operations on the parentheses, you can add and subtract as needed.

$$\frac{1}{6} + \frac{8}{6} - \frac{3}{6} =$$

$$\frac{9}{6} - \frac{3}{6} = \frac{6}{6}$$

$$\frac{6}{6} = 1$$

7) The correct answer is D. We know that Mary has already gotten 80% of the money. However, the question is asking how much money she still needs: 100% − 80% = 20% = .20
Now do the multiplication: 650 × .20 = 130

8) The correct answer is C. Any non-zero number to the power of zero is equal to 1. $(6y)^0 = 1$

9) The correct answer is C. To answer this type of question, remember that $x^{-b} = \frac{1}{x^b}$

Therefore, $-5^{-2} = \frac{1}{-5^2} = \frac{1}{25}$

10) The correct answer is D. You need to find the relationships between the numbers provided in the chart in order to determine the missing value.

STEP 1 – Consider whether a relationship between the numbers on the first row of the table can be found based on addition or subtraction. Look at each of the sets of numbers on a line by line basis. On the first line, we have 2 in the left column and 4 in the right column. So, we can get to the value in the left column by adding 2.

STEP 2 – Try out the value calculated in step 1 for the next row of numbers: $4 + 2 \neq 16$

STEP 3 – If the relationship does not work for the second row of number we have to consider whether the relationship between the numbers is based on multiplication or division. Returning to row 1 of the table, we can determine that: $2 \times 2 = 4$

STEP 4 – Try this operation on the second row of numbers: $4 \times 2 \neq 16$

STEP 5 – Try to determine if any other relationship is possible. Since $2 \times 2 = 4$ on the first row of the table, we can also try multiplying each subsequent number by itself.

STEP 6 – Try this new relationship for the second and subsequent rows.

Row 2: $4 \times 4 = 16$

Row 4: $8 \times 8 = 64$

Row 5: $10 \times 10 = 100$

STEP 7 – Calculate the value missing from row 3.

Row 3: $6 \times 6 = 36$

11) The correct answer is C. To solve inequalities like this one, you should first solve the equation for x.

$x - 6 < 0$
$x - 6 + 6 < 0 + 6$
$x < 6$

Now solve for y by replacing x with its value.
$y < x + 12$
$y < 6 + 12$
$y < 18$

12) The correct answer is B. Our problem asked for the solution of $2 + y < -8$. Isolate y in order to solve the problem.

$2 + y < -8$
$2 - 2 + y < -8 - 2$
$y < -10$

13) The correct answer is C. The first point on the graph lies at $x = 10$, so we can eliminate answer choices A and B. The point for the y coordinate that corresponds to $x = 10$ is 63 not 68, so we can eliminate answer choice D.

14) The correct answer is A. To divide, invert the second fraction and then multiply as shown.
$$\frac{x}{5} \div \frac{9}{y} = \frac{x}{5} \times \frac{y}{9} = \frac{x \times y}{5 \times 9} = \frac{xy}{45}$$

15) The correct answer is A. The ratio of defective chips to functioning chips is 1 to 20. So, the defective chips form one group and the functioning chips form another group. Therefore, the total data set can be divided into groups of 21. Accordingly, $1/21$ of the chips will be defective. The factory produced 11,235 chips last week, so we calculate as follows: $11{,}235 \times 1/21 = 535$

16) The correct answer is D. Calculate the discount: $120 × 12.5% = $15 discount. Then subtract the discount from the original price to determine the sales price: $120 − $15 = $105

17) The correct answer is D. Get the integers to one side of the equation first of all.

$$\frac{1}{5}x + 3 = 5$$

$$\frac{1}{5}x + 3 - 3 = 5 - 3$$

$$\frac{1}{5}x = 2$$

Then multiply to eliminate the fraction and solve the problem.

$$\frac{1}{5}x \times 5 = 2 \times 5$$

$$x = 10$$

18) The correct answer is A. For algebraic equivalency questions like this, you can perform the operations on each of the answer choices to see which one is equivalent. Remember to be careful when performing multiplication on negative numbers inside parentheticals.
6 + 2(15 − x) =
6 + (2 × 15) + (2 × −x) =
6 + 30 − 2x =
36 − 2x

19) The correct answer is D. The formula for the area of a circle is: $\pi \times R^2$. First, we need to calculate the area of the larger circle: $\pi \times 2.4^2 = 5.76\pi$. Then calculate the area of the smaller inner circle: $\pi \times 1^2 = \pi$. We need to find the difference between half of each circle, so divide the area of each circle by 2 and then subtract.

$$(5.76\pi \div 2) - (\pi \div 2) = \frac{5.76\pi}{2} - \frac{\pi}{2} = \frac{4.76\pi}{2} = 2.38\pi$$

20) The correct answer is C. Calculate the area for each cupboard: 8 × 2 = 16 and 5 × 2 = 10. Therefore, the total area for both cupboards is 16 + 10 = 26. Then find the area for the entire kitchen: 8 × 12 = 96. Then deduct the cupboards from the total: 96 − 26 = 70

21) The correct answer is B. The two sides of the pen form a right angle, so we can use the Pythagorean Theorem to solve the problem: $\sqrt{3^2 + 4^2} = \sqrt{9 + 16} = \sqrt{25} = 5$

22) The correct answer is A. First, we need to find the circumference of the semicircle on the left side of the figure. The width of the rectangle of 10 inches forms the diameter of the semicircle, so the circumference of an entire circle with a diameter of 10 inches would be 10π inches. We need the circumference for a semicircle only, which is half of a circle, so we need to divide the circumference by 2: $10\pi \div 2 = 5\pi$. Since the right side of the figure is an equilateral triangle, the two sides of the triangle have the same length as the width of the rectangle, so they are 10 inches each. Finally, you need to add up the lengths of all of the sides to get the answer:
18 + 18 + 10 + 10 + 5π = 56 + 5π inches

23) The correct answer is D. The area of a circle is always π times the radius squared. Therefore, the area of circle A is: $3^2\pi = 9\pi$. Since the circles are internally tangent, the radius of circle B is calculated by taking the radius of circle A times 2. In other words, the diameter of circle A is the radius of circle B. Therefore, the radius of circle B is 3 × 2 = 6 and the area of circle B is $6^2\pi = 36\pi$. To calculate the remaining area of circle B, we subtract as follows: $36\pi - 9\pi = 27\pi$

24) The correct answer is A. The perimeter of a rectangle is equal to two times the length plus two times the width. We can express this concept as an equation: P = 2L + 2W
Now set up formulas for the perimeters both before and after the increase.

STEP 1 – Before the increase:
P = 2L + 2W
48 = 2L + 2W
48 ÷ 2 = (2L + 2W) ÷ 2
24 = L + W
24 – W = L + W – W
24 – W = L

STEP 2 – After the increase (width is increased by 5 and length is doubled):
P = 2L + 2W
92 = (2×2)L + 2(W + 5)
92 = 4L + 2W + 10
92 – 10 = 4L + 2W + 10 – 10
82 = 4L + 2W

Then solve by substitution. In this case, we substitute 24 – W (which we calculated in the "before" equation in step 1) for L in the "after" equation calculated in step 2, in order to solve for W.
82 = 4L + 2W
82 = 4(24 – W) + 2W
82 = 96 – 4W + 2W
82 – 96 = 96 – 96 – 4W + 2W
–14 = –2W
7 = W

Then substitute the value for W in order to solve for L.
24 – W = L
24 – 7 = L
17 = L

25) The correct answer is C. Count how many blocks lie along the outer edges of the shaded area in order to get your result. Top boundary = 4 feet. Left side boundary = 5 feet. Bottom boundary = 3 feet. Right boundary = 6 feet (Don't forget to count the piece shaped like the upside-down "L" on the right.) Then add these amounts to get your result: 4 + 5 + 3 + 6 = 18 feet

26) The correct answer is B. He owns 26 yachts and needs 6 feet 10 inches of rope for each one. Convert the feet and inches measurement to inches. 6 feet 10 inches = (6 × 12) + 10 inches = 72 + 10 = 82 inches. Then multiply by the number of items: 26 × 82 = 2132 inches of rope needed. Then convert back to feet and inches: 2132 inches ÷ 12 = 177 feet 8 inches

27) The correct answer is D. Measure the length along the top and bottom of the frame, as well as the length of both sides in order to get the basic perimeter. Top = 4 inches. Bottom = 4 inches. Left side = 6 inches. Right side = 6 inches. Total perimeter: 4 + 4 + 6 + 6 = 20 inches. Now add in the 4 extra inches for the four corners: 20 + 4 = 24 inches.

28) The correct answer is B. Questions about measuring time and distance are non-statistical questions because they have only one possible answer. "How many customers go to that coffee shop on Saturdays?" is a statistical question because the amount of customers in the shop will vary from one weekend to the next.

29) The correct answer is D. The proposed course of action affects the residents of Buford, so the result must represent these residents as much as possible. The most representative result would therefore be achieved by selecting participants for a survey from a list of all of the citizens living in Buford. Answers A and B may represent people living in other areas who are merely driving near Buford. Answer C would fail to represent residents of Buford who do not drive.

30) The correct answer is D. We have 54 cards in the deck (13 × 4 = 52). We have taken out two spades, one heart, and a club, thereby removing 4 cards. So, the available data set is 48 (52 – 4 = 48). The desired outcome is drawing a heart. We have 13 hearts to begin with and one has been removed, so there are 12 hearts left. So, the probability of drawing a heart is $^{12}/_{48} = ^{1}/_{4}$

31) The correct answer is B. Our data set is: 2.5, 9.4, 3.1, 1.7, 3.2, 8.2, 4.5, 6.4, 7.8. First, put the numbers in ascending order: 1.7, 2.5, 3.1, 3.2, 4.5, 6.4, 7.8, 8.2, 9.4. The median is the number in the middle of the set: 1.7, 2.5, 3.1, 3.2, **4.5**, 6.4, 7.8, 8.2, 9.4

32) The correct answer is D. The last 40 minutes of the journey begin at the 80 minute mark and end at the 120 minute mark. The line for 80 minutes is at 520 miles and the line for 120 minutes is at 780 miles, so the plane has traveled 260 miles (780 – 520 = 260) in the last 40 minutes.

33) The correct answer is B. The event is defined as the chance of rain. In terms of probabilities, we know that there are 365 days in non-leap years, so this goes in the denominator. The chance of rain goes in the numerator: $^{45}/_{365}$ = 12.3288%. However, the forecast is correct only 90% of the time: 12.3288% × 90% = 11.0959%

34) The correct answer is D. The median is the number that is halfway through the set. Our data set is: 19, 20, 20. 15, 21, 18, 20, 23, 22, 15, 12, 23, 9, 18, 17. First, put the numbers in ascending order: 9, 12, 15, 15, 17, 18, 18, 19, 20, 20, 20, 21, 22, 23, 23. We have 15 numbers, so the 8th number in the set is halfway and is therefore the median:
9, 12, 15, 15, 17, 18, 18, **19**, 20, 20, 20, 21, 22, 23, 23

35) The correct answer is A. The outcome of an earlier roll does not affect the outcome of the next roll. When rolling a pair of dice, the possibility of an odd number is always $1/2$, just as the possibility of an even number is always $1/2$. We can prove this mathematically by looking at the possible outcomes:

1,1 1,2 1,3 1,4 1,5 1,6
2,1 2,2 2,3 2,4 2,5 2,6
3,1 3,2 3,3 3,4 3,5 3,6
4,1 4,2 4,3 4,4 4,5 4,6
5,1 5,2 5,3 5,4 5,5 5,6
6,1 6,2 6,3 6,4 6,5 6,6

The odd number combinations are highlighted:

1,1 **1,2** 1,3 **1,4** 1,5 **1,6**
2,1 2,2 **2,3** 2,4 **2,5** 2,6
3,1 **3,2** 3,3 **3,4** 3,5 **3,6**
4,1 4,2 **4,3** 4,4 **4,5** 4,6
5,1 **5,2** 5,3 **5,4** 5,5 **5,6**
6,1 6,2 **6,3** 6,4 **6,5** 6,6

So, we can see that an odd number will be rolled half of the time.

36) The correct answer is A. We want to see the percentage that each class is to the total enrollment for the high school. A pie chart is best for this purpose. For example, if there are four classes, the pie would have "slices" representing the enrollment in each class at the beginning of the year.

37) The correct answer is D. For questions like this, you will recall that the points could lie on one continuous strait path like a line. Alternatively, the points could be laid out more like a triangle. However, the distance between points will always be greater when the points are linear. If the points are linear then the maximum distance will be calculated as follows:

12 miles + 18 miles = 30 miles

38) The correct answer is A. The problem tells us that A is 3 times B, and B is 3 more than 6 times C. So, we need to create equations based on this information.
B is 3 more than 6 times C: B = 6C + 3
A is 3 times B: A = 3B
Since B = 6C + 3, we can substitute 6C + 3 for B in the second equation as follows:
A = 3B
A = 3(6C + 3)

A = 18C + 9

So, A is 9 more than 18 times C.

39) The correct answer is A. The total amount that Toby has to pay is represented by C. He is paying D dollars immediately, so we can determine the remaining amount that he owes by deducting his down payment from the total. So, the remaining amount owing is represented by the equation: C – D

We have to divide the remaining amount owing by the number of months (M) to get the monthly payment (P): P = (C – D) ÷ M = $\frac{C-D}{M}$

40) The correct answer is A. The problem tells us that the morning flight had 52 passengers more than the evening flight, and there were 540 passengers in total on the two flights that day. First of all, we need to deduct the difference from the total: 540 – 52 = 488; In other words, there were 488 passengers on both flights combined, plus the 52 additional passengers on the morning flight. Now divide this result by 2 to allocate an amount of passengers to each flight: 488 ÷ 2 = 244 passengers on the evening flight. (Had the question asked you for the amount of passengers on the morning flight, you would have had to add back the amount of additional passengers to find the total amount of passengers for the morning flight: 244 + 52 = 296 passengers on the morning flight)

41) The correct answer is B. At the beginning of January, there are 300 students, but 5% of the students leave during the month, so we have 95% left at the end of the month: 300 × 95% = 285. Then 15 students join on the last day of the month, so we add that back in to get to the total at the end of January: 285 + 15 = 300. If this pattern continues, there will always be 300 students in the academy at the end of any month.

42) The correct answer is C. Round each item up or down to the nearest dollar. We have the values $1.99, $20.05, $98.75, $0.35, 103.62. So we add as follows: 2 + 20 + 99 + 0 + 104 = 225

43) The correct answer is B. Our series was: $\frac{1}{4}, \frac{5}{8}, \frac{9}{12}, \frac{13}{16}$

So, the numerator and denominator increase by 4 each time.

$$\frac{13+4}{16+4} = \frac{17}{20}$$

44) The correct answer is B. Our facts were: 5 more than 4 times the number x is equal to the number y. Build up your equation based on each part of the problem.

4 times the number x: 4x

5 more than 4 times the number x: 4x + 5

5 more than 4 times the number x is equal to the number y: : 4x + 5 = y

45) The correct answer is C. Paul won't sit next to Maria, so answers A and B are incorrect. Jeri has to sit next to Sue, so answer D is incorrect. The arrangement "Paul, Sue, Jeri, Tom, Maria, You" is the only one in which Tom doesn't sit next to Paul, Paul doesn't sit next to Maria, and Jeri is next to Sue.

www.ingramcontent.com/pod-product-compliance
Lightning Source LLC
Chambersburg PA
CBHW081351080526
44588CB00016B/2447